Television and Cultural Crisis

An Analysis of Transnational Television in India

Mira K. Desai
Binod C. Agrawal

CONCEPT PUBLISHING COMPANY, NEW DELHI-110 059

ISBN-13: 978-81-8069-609-1 ISBN-10: 81-8069-609-X

First Published 2009

Published and Printed by

Concept
Since 1974

Concept Publishing Company
Regd. Office:
A/15-16, Commercial Block, Mohan Garden
New Delhi-110059 (India)
Phones: 25351460, 25351794, *Fax*: 091-11-25357109
Email: publishing@conceptpub.com,
website: www.conceptpub.com

Editorial Office:
H-13, Bali Nagar, New Delhi-110 015, India.

Cataloging in Publication Data--*Courtesy:* D.K. Agencies (P) Ltd. <docinfo@dkagencies.com>

Desai, Mira K.
Television and cultural crisis : an analysis of transnational television in India / Mira K. Desai, Binod C. Agrawal.
p. cm.
Originally presented as the first author's Thesis (Ph. D.)--2003.
Includes bibliographical references and index.
ISBN 13: 9788180696091 ISBN 10: 818069609X

1. Television broadcasting--Social aspects--India. 2. Mass media and culture--India. I. Agrawal, Binod C. (Binod Chand), 1942-, joint author. II. Title.

DDC 302.23450954 22

Foreword

Television in India arrived half a century ago as a modest effort of Central Government. Its tremendous expansion has enabled India to become one of the world's largest television nation in providing information, education and entertainment. Today, television access and reach in India is unparallel to any other country having penetrated in large majority of urban and rural homes. Video technology fueled its growth and indigenous expansion of satellite communication provided distance neutral access to viewers regardless of where they live. The television expansion has been uninterrupted in spite of major socio-cultural upheavals and the resultant influence on the viewers across the country, as claimed by the experts, is thought provoking.

In this book, Desai and Agrawal have provided an analysis of television influences on the urban viewers. The major question raised relates to the influences of television viewing of English language programmes of foreign origin which the authors have refeired to as "*transnational television*". Similarities and differences of influences of television viewing and other media have been examined among the two linguistic urban groups - the Gujarati and Marathi speaking households of metropolitan Mumbai. Influence of television viewing has been examined on the cultural values of the viewers.

The television viewing in Mumbai households, like elsewhere in the country, was found to be "*family centric*". The authors have argued that television helped in reinforcing the values of family without making any major dents in it. At the same time family interaction was found to be the most influenced as a result of family television viewing. Eating habits was least influenced in spite of continuous television advertisements of new or western food. Marathi viewers reported increased family interaction because of television as compared to Gujarati viewers. Not so surprising, Mumbai viewers, in spite of being in cosmopolitan surrounding, were found to be largely attracted to the Indian language television and not more than 10 percent cared to view English language television.

Viewers regardless of their language preferences and economic status preferred 'local or localized' entertainment. Gujarati and Marathi speaking viewers depended on locally produced television content and quite a few watched local cable channels as well. 'English channel' viewers had studied in English medium schools, were speaking English at home among family members, and consuming other forms of media in English.

Younger viewers in 20 to 30 years of age reported that television had influenced them psychologically (personal thinking) and materially (desire for buying new products and goods and enhanced consciousness about physical self) compared to older viewers in 40 to 50 years of age group. It was heartening to note that the impact of television created the need for buying material things among not so well to do viewers, while rich viewers used television largely as source of information.

I recommend this book for reading to both media scholars and television channel executives to correct their misconceptions about the Indian television viewers and their preferences.

Subhash Chandra
Chairman
Zee Entertainment Enterprises Limited
And Essel Group
135 Continental Building
Dr. Annie Besant Road,
Worli, Mumbai 400 018

Preface

The idea of writing the book on "impact" of television on the Indian Civilisation was nurtured by both of us while Mira finished her Ph.D. thesis under the guidance of Binod C. Agrawal in 2003. The research question raised by us was how to test the validity of the position taken by communication researchers, policy makers and political leaders that the Western transnational television, especially in English produced in Europe and America was eroding the cultural values of traditional South Asia. The concern was that their cultural values are being replaced by Western values due to onslaught of transnational television.

The question assumed even greater importance keeping in mind that more than two centuries of direct British contact with South Asian countries have had limited impact. The protagonists of communication revolution quickly concluded that the transnational television has brought about a new form of "media colonialism" leading to a "cultural-media imperialism" in which no direct brutal force was being used. The communication experts argued that television invaded a large number of South Asian homes and will exert greater influence in the cultural lives of these viewers. This book is the outcome of efforts made after 2003 based on the further analysis carried out on the data collected during 2002.

This book examines extend to which private satellite television has brought about "cultural-media imperialism" in South Asia in general and India in particular. The analysis focuses on the post-economic liberalisation period of 1990, which coincided with the rapid expansion, and deregulation of broadcasting in the countries like India, Bangladesh, Pakistan, Sri Lanka. Attempts have been made to present the impact analysis of transnational television viewing of both

English and South Asian language programmes. Two dominant television programme genre that continue to grow: are one, indigenusly produced television programmes in South Asian languages and secondly, English language programmes produced in Euro-American countries. The dominant cultural values of Western culture were selected to examine the extent to which the viewing of English language television programme has brought about change in the Indian viewers. These cultural values relate to values of "individualism" and "consumerism" dominant in the West as opposed to collectivism/familialism and austerity value orientations of East.

The study was carried out amongst two linguistic community households; Gujaratis, who speak Gujarati language and Marathis who speak Marathi. The sample for the study was drawn from India's most modern mega city Mumbai supposed to have adopted Western values. In Mumbai, both Marathi and Gujarati speakers are numerically dominant group. Among 400 selected households covering a total of 1007 Gujarati and 1015 Marathi members of both the sexes who owned Cable and Satellite connection and viewed transnational television in varying degrees through individual interviews of 400 viewers in 20 to 50 years of age.

Majority of the viewers in the sample had exposure to English or foreign television content though; very few reported viewing only English language foreign channels. High preference for 'local and localized' Indian language channels was reported. The small number of viewers who watched English or foreign channels spoke in English at home had prior exposure to English as medium of instruction at school and consumed "non-television media" in English.

The daily average time spent on media including television was three hours, of which one hour forty minutes was spent on television. Women spent little less time on television compared to men though both spent more or less same amount of time on media other than television. Less-educated viewers spent more time on "media including television" whereas better educated spent more time with "media other than television". Middle-aged (30 to 40 years of age) viewers spent least amount of time with media compared to those below 30 and above 40 years of age. It was found that as compare to Gujaratis, Marathis usage of media was higher in terms of time and variety.

Influence of English langauge transnational television on the cultural values could not be established. Transnational television viewers of upper economic status were found to be more individualistic and consumerist as compared to the others. Also television-viewing behaviour alone, as analysis revealed, could not explain behavioural change.

We conclude that transnational television, as earlier believed, may not bring about cultural-media imperialism in South Asia in general and India in particular even amongst the most "Westernised" segment of urban viewers. Transnational television consumption is driven by individual as well as collective variables within the Indian households. Though further in-depth studies to identify life-cycle usage of media, role of television in viewers' other media usage and the core strength of South Asian cultures that withstand media invasion, would be useful.

September 5, 2008

Mira K. Desai
Binod C. Agrawal

Acknowledgements

The project was conceived as it was only because of vision and conviction of Dr. Binod Agrawal who was on the mission to prove that Indian way of media consumption needed to be viewed from Indian angle and western theoretical premises may not explain the 'reality'. It was he, who contributed intellectually to this work and because of whom this book is seeing the light of the day.

Mumbai being the city as is, if the respondent would not have given time and energy to the project, the quality of work would not have been as it turned out to be. I am indebted to all the respondents for sparing time and effort to entertain me, an outsider, to their lives. Special mention is needed of all those: Anilamasi-Himanshumasa, Rajnibhai-Hemabhabhi, Alkaben, Jasmin-lali-ben, Meenabhabhi, Vishwa, Suhana and Jagdishbhai who helped me at various stages of data collection. Neela Dabir, Veena Karnik and Nanda Satope need special mention for their help in terms of language, Marathi.

Mr. Nitin Mittal agreed for publishing the manuscript without any explanation suggested his faith in the material.

My families—Papa, Ma, Anand and Kuntal besides my father-in-law and mother-in-law who stood by me when I worked on this research, contributed in their own ways. My husband Kapil is special, as being what he is, without whose silent support this book would not be in your hands, get all the credit for what I could do with this project and my life post 1996.

Mumbai
August 27, 2007

Mira K. Desai

Contents

List of Tables

Annexure-E

List of Figure, Graphs and Visuals

Figure

Graphs

Visuals

1

Introduction

Culture...the much debated term in the era of globalised world has various connotations. Culture as civilisation, culture as the worldview, culture as values, norms and artifacts, culture as patterns of products and activities, culture in form of day-to-day practices. Culture is the part of all forms of human interaction. Sociologists debate about association of terms society and culture, anthropologists about culture and civilisation. Since the late nineteenth century, anthropologists had adopted and adapted the term *culture* to a broader definition that they could apply to a wider variety of societies.

The earliest and the most quoted definition of culture is by Sir Edward B. Tylor (1871), "culture or civilization, taken in its wide ethnographic sense, is that complex whole which includes knowledge, belief, art, morals, law, custom, and any other capabilities and habits acquired by man as a member of society". The United Nations agency, UNESCO (2000), defined culture as the "set of distinctive spiritual, material, intellectual and emotional features of society or a social group, and that it encompasses, in addition to art and literature, lifestyles, ways of living together, value systems, traditions and beliefs".

In the twenty-first century, culture as institutions, technologies and ideologies has been critical area of research. Globalisation has also led to more avenues for cultural interaction, conflict, adaptation. Media has a very important role to play in this processes of cultural exchange. Though radio and films existed over decades leading to transmission of cultural products across geographical, political and social boundaries, post-1990s television has become the real common denominator.

Television...as a medium of mass communication is one of the contributions towards integration of technology, culture, commerce and politics. Television largely reflects the culture of the society through audio-visual codes. Television through verbal, non-verbal and symbolic codes project elements of material (artifacts, food, clothing, sports etc.), abstract ideas (ethos, values, attitudes, beliefs etc.), politics, kinship and religion. Television as a cultural product reflects the cultural values of their producers and images of the social reality in which they are produced.

After the first public broadcast in 1926, by the end of twentieth century television has become instrumental in blurring geographical and political boundaries. Cable News Network (CNN) with its live telecast of fall of Berlin wall in 1989 and Gulf war in 1991 lead to uniform dissemination of American news broadcasting style across the globe.

Statistics show 73.4 per cent rise in television sets per 1000 persons around the world between 1980 and 1997. The growth is enormous in "developing countries (92 per cent) and marginal (2.2 per cent) in industrial countries" (UNESCO, 2000: 303). The rapid expansion of television across the globe can be credited to satellite broadcasting and advances in telecommunications. The marriage of cable and satellite technology provided a phenomenal increase in the availability of television channels and made global exchange of television content a day-to-day reality.

The popularity of the medium can be attributed to the fact that it reaches the household directly in a large part of the world. Viewing television is not merely an act of passive consumption but is "rather complex process of decoding cultural meanings" (Wang *et al.*, 2000:4).

Historically the phenomenon of unidirectional flow of information and cultural elements from the Euro-American countries to the countries of Africa, Asia and South America, has been subject of much academic and political debate. The debate assumed importance also due to the expressed fear of possible adverse effects of 'foreign' programmes on television viewers of the traditional civilizations like India. In the last decade and a half the debate about influence of transnational television on moral degradation and damage on the social fabric of Asian societies has been intensified. Further, it is believed that transnational television is increasingly popularizing western values to non-western societies of Asia and thus impacting local cultures.

Transnational television in the form of foreign programmes began appearing on Indian television in the late 1970s. Kumar (1999: 100) notes, "around 1980s the only sitcoms, soap operas, detective and other TV genres telecast were from British, United States or German television. The British sitcoms shown during those early years of the national network were very regional in their humour and accent...American serials like 'I Love Lucy', 'Star Trek', German 'Telematches' and detective serials like 'The Fox' introduced Indian viewers to the other kinds of foreign fare". In 1990 transnational television in form of Cable and Satellite Television (C&STV) had the nature and the scale of foreign programming on Indian television beyond comparison. Availability and access of transnational television for almost half of the Indian television viewers is continuous and non-interrupted. It has reached over 80 per cent of the television households in Mumbai and so is the case with other metropolitan cities in India and close to sixty per cent of television households in the country.

Lot has been written about the influence of western television programmes on Indian viewers, but little empirical data exist. Researchers have stated that "not much is known about consumption pattern of foreign *vis-à-vis* domestic products or the nature of impact from transnational media" (Wang *et al.*, 2000:12), and "local audience preference of domestic programmes to imported ones has not been proven when both are competing with one another at the same viewing time"; (Lee and Joe, 2000: 139). Researches have indicated that television-viewing influences value orientations of the viewers. Television's association with consumerism has been long established by several researchers (Wells 1972, Vilanilam 1989, Cashmore 1994, Dixit 1998, Chandra 2000). But there is little or no empirical confirmation about influence of Western or transnational television programming on the family-oriented values of Asian television viewers.

The book aims to:

- understand the nature of influence of transnational television viewing on Gujarati and Marathi speaking households in Mumbai;
- examine the perceptions of the viewers regarding implications of transnational television programmes on their lives and on the society; and
- explore the nature of influence of transnational television viewing with specific reference to the values of individualism and consumerism.

The reason of choosing values of individualism and consumerism is both, to an extent, are representations of capitalist world value system. The cry for the cultural crisis and erosion of collective and altruistic value orientations of Asian societies by the Western television programming has been the concern of research for a fairly long time. The theoretical premise of cultural imperialism reflected the popular 'cry' for cultural degradation and moral erosion of eastern societies due to massive onslaught of western media content.

Cultural-Media Imperialism...premise has been empirically tested through present research. Historically 1980s witnessed regional exchanges of television programmes within and across countries. UNESCO led the debate of restructuring of international information and communication order. As a result, studies on communication flows across the globe were initiated (Nordenstreng and Varis 1974, Guback and Varis 1982, Varis 1985). The studies pointed out that "it is mainly with reference to the one-way, non-reciprocal flow of information and cultural influence in international media environment, which embodies cultural domination of the 'third world' by the West" (Wang, 1997:15-19).

In 1969, Herbert Schiller was the first to document the way in which the American military-industrial complex strived to maintain a worldwide electronic communication hegemony and cultural domination. Schiller (1976: 9) described 'cultural imperialism' as "the sum processes by which a society is brought into the modern world system and how its dominating stratum is attracted, pressured, forced and sometimes bribed into shaping social institutions to correspond to, or even promote, the values and structures of the dominating centre of the system." Schiller considered American approach of promoting the interests of USA, directly through the channels of advertising and indirectly through the display of lifestyles. Lee (1980) extended Schiller's concept of cultural imperialism to media imperialism and remarked, "media imperialism [is] as composite of programme flows, ownership transfers of broadcast systems and promotion of capitalist worldviews/ lifestyles...the term cultural imperialism was preferred by those with Marxist bent to imply wholesale domination of which media are mere symptoms" (Lee, 1980).

The terms: cultural imperialism and media imperialism too has been variedly defined. Sreberny-Mohammadi (1997:51) is of the opinion that "the driving force behind the propagation of cultural imperialism is the media. It is the vehicle, which delivers new thoughts and ideas. Media imperialism is central to cultural imperialism". Contrary to that Tomlinson (1997) argues that the media is not the totality of cultural experience but is merely the 'mediation' of capitalist modernity, which is a form of global capitalist imperialism.

Communication scholars who treat the two terms "cultural imperialism" and "media imperialism", as synonymous bring into question the centrality of media in the claims of cultural imperialism. In this respect, culture and media are viewed as one and same in which, media is considered as all encompassing as culture.

Cultural-media imperialism theories had predicted that the importation of television programmes would lead to demise of local cultures and cultural industries without considering complex patterns of television production, distribution, and consumption. Present discourse of cultural-media imperialism refers to concepts like 'localization', 'globalization', 'counter-flow', 'local resistance', 'cultural proximity', 'cultural/economic capital' (Wang 1997, Thussu 1998, Chadha and Kavoori 2000, Straubhaar 2000).

Transnational Television

Transnational television as a term has conceptual and operational differences with other terms like international and multinational. Modelski (1979: 2) defines transnational as "corporations that are nationally-based and attend their business to other countries; whereas 'multinational' or 'international' signifies international allocation of resources and production, and most importantly, international management and ownership". In the twenty-first century 'global media' has taken variety of forms "ranging from direct transmission or distribution of media channels, to import of formats and genres of foreign origin, international news items, and miscellaneous content such as sporting events, advertising and pictures that have a foreign reference or origin" (McQuail, 2000: 220).

Transnational has been defined in variety of ways. Reddi defines transnational television as "broadcast services that are essentially satellite-based, with production of content in one country, transmission from a second country, and addressed to viewers in a third" (Reddi, 2000: 69). Reddi's definition does not refer to the cultural dimension of message production, transmission, and consumption. Joshi (1998: 6) defines transnational television programmes in variety of ways. He refers to 'foreign' programmes telecast on Indian television channels as well as foreign channels or channels of Indian origin beamed through satellites, programmes of neighbouring countries received indirectly by cable networks, and cable-casting of pirated or other material from foreign countries.

Transnational television has also been referred to as "international satellite broadcasting", "television without border", "cross border television", "transborder television", "global television" or "satellite television". Transnational television here means television programming from 'other countries'. In India transnational television is distributed through cable connection and therefore is known as Cable and Satellite Television (C&STV), which essentially meant television broadcasting of public and private satellite television channels.

Transnational Television Viewers

Mass media research, growing out of propaganda studies in First World War, assumed that media had direct influence over their viewers...(such a view) has now faded and been replaced by the complex transactional perspective of social psychology (Nord, 1983). Audience reception studies in 1980s indicated that viewers were important decision-makers in transnational television reception. It was realized that "language, culture and class can both facilitate and provide barriers against cross-border flow of television programmes" (Wang *et al.*, 2000:13).

In 1990s audience preferences led to change in programming strategies of transnational broadcasters. It was concluded that all over the world television viewers prefer local, 'culturally proximate' programming as against transnational television programmes (Kottak, 1990; Straubhaar, 2000; Lee, 2000).

Present research attempted to examine if television viewers get influenced by transnational television as assumed in the earlier studies of transnational television flow. It takes the view that television viewers are 'active viewers' (McQuail, 1997), 'autonomous reader' (Hagen and Wasko, 2000), having 'polysemy' of message reading, 'cultural capital', and 'cultural eclecticism' (Oliveira, 1986; Fiske, 1987; Goonsekera, 1987; Tomlinson, 1991; Hamelink, 1994; Golding and Harris, 1997; Mankekar, 1999; Straubhaar, 2000; Wang, 2000; Eko, 2001) impacting their viewing and they resist and block such influences. Instead of researching flow of programming, "by the 1990s several scholars of globalisation had begun to address consumption and the formation of transnational consumption communities, as key issues and foci for research" (Griffin, 2002:27).

Language and Television

In the formation of television markets within or across national boundaries, almost all studies to date agree on the importance of language in creating television markets. This has been confirmed by economic studies of market... At the individual audience level, competence, ability to at least understand the language of a broadcast is an important ingredient in viewers' selection of programme and their enjoyment of it (Straubhaar, 2000: 205-6).

In Asian context, plurality of languages, dialects and cultures are critical issues in television programme distribution and consumption. Straubhaar, Campbell and Cahoon (1995) reported " 'regionalisation' as a more significant phenomenon than the idea of globalisation, *per se* for television into multi-country markets linked by geography, language and culture". They commented that these might more accurately be called cultural-linguistic markets, rather than regional markets, since not all these linked populations, markets and cultures are geographically contiguous.

But the danger in treating culture and language as another set of powerful determining factors in communication research thus undermines the importance of other factors. Servaes and Lie (2001: 21) remarked, "in fact, no single factor, nor a group of factors, can fully explain what has, is or will take place". It appears that the Inter-linkages of communication, culture and language are complex.

STAR (Satellite Television for Asian Region) TV, key player in private satellite television channel market divided its broadcast to Asia into two; Northern beam covering eastern Asia including China, Hong Kong and Taiwan and the Southern beam to cover South Asia including Indian subcontinent and Middle East. Northern beam had English programmes either dubbed or subtitled into Mandarin while Southern beam was in Hindi or Arabic. From October 1996 the scenario changed. STAR Plus Indianized its programming and lot of content was produced exclusively for the Indian markets in Hindi language or at times in *Hinglish* (combination of Hindi and English). Today, STAR TV operates 40 channels in eight languages in 53 countries (http://www.startv.com).

Sinclair (1998) comments that the STAR's strategy of 'going local' shows how much language and culture have emerged a 'tangible markets'. Sinclair (1998:222) further remarked, "even more striking is the conclusion that cultural and linguistic differences have in themselves proved to be much more substantial barriers to foreign programming than the theory of cultural imperialism would have predicted, and that the global service providers have had to accede to the demand for the popular programming in the languages of the region" (Sinclair, 1998:222).

Many of the Indian and foreign channels have opted for bilingualism or even expansion in other Indian languages. Zee TV now has over 15 channels in Indian languages at the same time has international channels like ZEE MGM, ZEE English and lately fashion channel Trenz. Zee is a case of 'going global' unlike STAR which went 'local'.

Thussu while examining the use of *'Hinglish'* (Hindi + English) on Indian television comments that its use "has been the contributory factor in the expansion of Indian television outside the borders of the country" (Thussu, 2000: 308). And at the same time "increasing emphasis on entertainment-led *Hinglish* television may be cause of celebration for post-modern view of culturally hybrid, globalised world resultant linguistic and intellectual confusion may be trivialization of vital public concerns" (Thussu, 2000: 308).

French and Richards (2000) complied number of studies about Singapore, Korea, China, Sri Lanka, Pakistan, and Indonesia detailing contemporary Asian examples of nature and importance of language in transborder television broadcasting. Pakistan has 66 languages and dialects of which ten are national languages and there is two-hour slot for language programming on channel-2. Sri Lankan television language use is "indicative of measure of tension between the state, nationhood and commercial imperative", as Sinhala is the language of state broadcaster and English is that of the Private channels. Singapore has four official languages—

Mandarin, Malay, Tamil and English. Mandarin was heavily watched and since 1995 *Singlish* entered on Singapore television, which is a combination of Mandarin and English.

Contrary to Asian example Latin American case of prevalence of Spanish language in nearly every nation, even with significant exception of Portuguese-speaking Brazil, had led to unification of North as well as South and Central America, and even Iberian nations of Europe. Cunningham, Sinclair and Jacha (1996) considers the possibility that "large numbers of channels allow regions defined by common cultural and linguistic ties to reach beyond physical space and engage in multidirectional dialogue". Brazil television TV Globo exports television shows to 128 countries so as Zee TV to 80 countries.

This book attempts to answer some of the questions of transnational television consumption among two linguistic communities—Gujaratis and Marathis residing in metropolitan environment of Mumbai. The reason for choosing linguistic group as focus of research was the understanding that language is a crucial part of culture. Hence, to explore the association of cultural values of individualism and consumerism with transnational television viewing among linguistic communities lead to evidences about influence of transnational television content on culture. Besides, though geographically located in the context of Mumbai, linguistic community is viewed as socio-cultural behaviour located within the cosmopolitan environment.

Language, Values and Culture

There are about 6000 languages being spoken across the globe. Language is the most recognizable part of any culture. "Language is defined as a learned system of arbitrary vocal symbols by means of which human beings as members of a society, interact and communicate in terms of their total culture" (Trager, 1972: 10).

"The impact of culture on biological existence, sociological relationships, spiritual convictions and aesthetic presentation is inevitable. And these elements refer to the culture of that society. Cultural profile of a society or nation is composite of two distinct elements: (i) the unique beliefs, values, and customs subscribed to by members of specific subcultures; and (ii) the central or core cultural elements that are shared by most of the population, regardless of specific subcultural memberships. Subcultural divisions are based on variety of socio-cutural and demographic variables, such as nationality, religion, geographic locality, race, age, sex, even working status" (Schiffman and Kanuk, 1999:440).

"While culture is concerned with the various ways we make sense of world, meanings are not simply 'out there' waiting for us to grasp them, they are generated through signs, most notably language...Language is taken to be at the heart of culture and identity for two central and related reasons. Language is privileged medium in which cultural meanings are formed and communicated. Language is the means and medium through which we form knowledge about ourselves and social world" (Barker, 2000:11). In the decade of 1960s Anthropologist linguistic Edward Sapir demonstrated the determinative effect of language on culture and worldview and argued that culture is largely psychological.

"Harry Triandis, Hazel Markus and Shinobo Kitayama, all social psychologists brought culture into psychology by arguing that shared values of social groups play key role in individual's cognitive, emotional and social functioning" (cited in Kitayama *et al.*, 1993: 4). Values have "a broad tendency to prefer certain states of affairs over others...it has both intensity and direction" (Hofstede, 1984:18).

As evident language, culture and values are interconnected and language and values can be considered as sub-set of culture. Values unlike attitudes are integral part of any culture as attitudes may be individual traits but values are collective identities.

Values of Individualism and Consumerism

Traditional institutions such as family, religious congregation and school form a foundation for the value system in that culture. Values are an attribute of individuals as well as collectivities; culture presupposes collectivity.

The value of 'individualism' means the importance of the individual as against community, collective, group and family (Kazdin, 2000; Triandis, 1996; Hui and Triandis, 1986; Britannica, 1983). Consumerism is diagonally opposed to the collective-responsibility and gives importance to individual consumption for personal pleasures. Consumerism essentially means "victory of capitalism as it denotes materialism...the acquisition orientation of individuals towards purchasing products" (Singhal and Rogers, 2001; Bearden *et al.*, 1993) "capitalism emphasized individualism which is pressuring the existing set-up family, which by and large demands group identity and group solidarity, and the purpose of doing it is market" (Sarkar, 1997: 265).

India, Languages, Television

India is a land of 5000 years old civilization and 50+ years old democracy. India is unique in more than one way. India, a union of 28 states and seven union territories, is a sovereign socialist secular democratic republic. It houses geographically, geologically, climatically, culturally, linguistically diversities. Present India is young (65 per cent below 30 years of age), better literate than earlier (decline in absolute number of illiterates from 1991 to 2001), populated (density 324 per square kilometre), rural (72 per cent) nation. It is a land filled with contradictions yet united by composite policy framework.

Most of the media in India are as old as in the world. Print, films, transnational television, Internet all began in India as it happened globally. Press, film, radio and transnational television in India owe it to private players whereas television and Internet took off under the government leadership. Indian media mosaic is as diverse as its society having varied ownerships, linguistic markets and complex distribution mechanisms.

There are five cultural regions in India: North, East, South, West, and North-east. There are six ethnic groups and five religious groups. According to 1961 Census Survey of India, there were 1652 mother tongues spoken in India of which 1549 were native to India and 33 languages are spoken by people numbering over a lakh. As per 1991 census there were a total of 114 languages and 216 mother tongues, 18 scheduled languages and 96 not specified in the schedule.

Today there are 22 officially recognized languages in India, 15 originally in the Constitution of India, followed by three and four languages added by amendments in 1992 and 2003 respectively. Hindi is the official language of the Union and English continues as second language for official work. Political struggles for separate state on linguistic grounds, hate-wave for Hindi in southern India, distinct north-south linguistic divide reflecting in film, television and publishing sectors are the highlights of language in India.

Vijayanunni (1999) reported that nearly 20 per cent of the population in India was bilingual while just over seven per cent were trilingual. As far as Gujaratis and Marathis are concerned, Government of India (1999) reported that for 24 per cent Marathis and 22 per cent Gujaratis, Hindi was a second or third language for communication.

Cable Waves (1998: 19) reported "among the home in urban India, 41 per cent speak a language at home, which remains largely unrepresented in the television programming. This includes Gujarati (6.4 per cent of homes), Bengali (7.7 per cent), Marathi (10.2 per cent) and Punjabi (3.4 per cent). Interestingly English is spoken in just 0.2 per cent of the urban homes a majority of which are in 23 largest cities". The situation changed in 1999 when many regional language channels were launched by private broadcasters as well as by Doordarshan. Annual Report of Doordarshan (2002) listed 15 Indian language channels available to television viewers in India of which 11 languages were covered by private satellite channels also.

What Zee TV is in mainly northern and western part of India, southern India saw launch of Sun TV (Tamil), Asianet (Malayalam), Eenadu TV (Telugu) and many more. English and Hindi are the dominant languages on Indian television scenario followed by Southern Indian languages.

Post-2000 Indian television scenario is overcrowded with channels for specialised content, segmented audiences and multiple languages.

Following pages elaborate about the role of television in Indian households, nature of television usage and its consequences as experienced by the viewers.

2

Television, Culture and Cultural-Media Imperialism

Television as a medium of mass communication is unique in more than one way. It is a cultural agent, 'provocateur and circulator of popular culture', 'social cement' and 'commodity in capitalist economy'. It has various functions: topic of social conversation, provider of family audience to advertisers, mirror of mass culture of a particular society through its codes of audio visual content and programme scheduling. Following features of television makes its unique.

1. Television reaches directly into the household and is most often statically positioned unlike radio. This leads viewer to the medium within the proximity of the household habitually.
2. After initial expenditure on acquiring the television set or now the cable connection to access transnational television, the renewable expense is only in form of cable subscription or electricity which is not direct as in case of newspapers or film viewing in cinema halls.
3. In spite of providing facility of individual consumption, it is a social medium which creates multiple meanings for its consumers unlike radio where experiences might be different but the meanings may remain same.
4. It is a medium that converges other mass media like radio, newspapers as well as films or Internet.
5. Its smaller physical size makes the viewer feel in control and the quality of image, environment and context of consumption makes it special unlike films.

Television reflects the culture in form of clothing, sets, makeup, jewellery, language, locations and so on. Apart from that television creates its own culture. Television as an integral part of culture of a given society, "is also dependent on the existing cultural values and patterns of social relationships in that society" (Goonasekera, 1987:9).

Television and Values

Researchers researching television content in Latin America have found that television programmes imported from the United States depict consumer society, plentiful in material well-being. These programmes show societies in which the central values are those of selfish individualism and intense competitiveness; where success and materialism are measured in terms of material possessions and where winning is everything (Sengupta and Frith, 1997: 3).

Association of television and values has been studied both from media content as well as media audience perspectives. But researchers have remarked that the framework of transnational telecasting is "encouraging commercial transactions rather than cultural ones" (Gupta, 1998: 84) and reflects "the demands of the *nouveau riche sexist* and sadistic violence, glorification of the ruthless go better, lure of the latest material gadgets are almost idealized by the emerging software linked with commercial agencies" (Joshi, 1985: 57). Though latter evidences suggested that cultural factors dictated television content rather strongly leading to shift in commercial transactions.

Cashmore (1994) argues that "the single most important danger posed by television is that it encourages people to spend too much. As the central apparatus of consumer society, television's success is measured not in terms of whether we enjoy programmes, but how much we spend as a result of watching them...This broadcast model tantalizes us with a lifestyle that is glamorous, instantly gratifying, bigger, better, and available to us; we can buy it. Television transforms culture from within, encouraging a shift from the work ethic to the consumption ethic. It is also argued that television fascinates because it embodies the culture it depicts. Television sets are brought, sold, upgraded, and supplemented, just like any other commodity. Our language echoes the language on television; the way we dress reflects television images. Television may be the most significant artifact of the century in its ability to give shape to and inform daily experience".

Sengupta and Frith (1997) in their content analysis of cultural values expressed in Multi-National Companies (MNC) commercials in comparison with the commercials of Indian Owned Businesses (IOB) observed that "MNC ads favour values such as modernity, cleanliness and technology (with more frequent use of Western models, settings, music and women in Western clothing)"(1997: 12-13).

Various researchers have studied and proved that television viewing can influence value orientations of the viewers. Murray and Kippax (1980: 619) noted, "television content can and does influence the viewer at various levels from changing attitudes to modifying interpersonal behaviour. television does act as a socializing agent". Wober and Gunter (1988: 220) reported about number of studies, which indicated that "viewers perceptions, beliefs and attitudes often relate more closely to viewing of particular kinds of programmes than to television viewing in total". Referring to the difficulty to summarize the role of television in the lives of adults, Murray and Kippax remarked that "the use that is made of television on the attitudes, values and behaviour of the viewers varies, to some extent, across cultures, and to a far greater extent across age, sex and social role status within a given culture" (1980: 619).

Many national as well as international studies about television viewing and values have concentrated on student population compared to other members of the society.

Kohn and Smart (1984) examined the effect of televised beer commercials on the beer consumption of young male drinkers. A videotaped indoor soccer game was shown to male students (sample size = 125). Different versions of the videotape included zero, 4, or 9 beer commercials. Fifty per cent of samples had immediate access to beer, and other half had access delayed by an hour. Logit analysis revealed that while exposure to the first few commercials increased consumption, continued exposure had no significant effect on total intake. Results suggest that experiments on alcohol advertising were likely to produce negative results when drinking was measured over a substantial period (e.g., over one hour), and positive results when measured for shorter periods.

Greeson and Williams (1986) in an investigation of the effects of Music TV (MTV) on adolescents' attitudes towards parental influence, premarital sex, violence, drug use, and MTV itself with seventh and tenth graders (sample size = 64) enrolled in a small-town school system in southwestern Ohio. The samples were shown MTV segments selected randomly. Following MTV viewing, samples responded to a Likert-type attitude survey concerning these issues. Results of analysis of variance show that: (a) the views of tenth graders reflected the views of their parents more closely than did those of seventh graders; (b) after viewing MTV, significantly more adolescents agreed that premarital sex was acceptable; and (c) no significant connection was found between MTV viewing and violence or drug use.

Kang and Morgan (1990) studied impact of US television in Korea using self-administered English questionnaire in April 1984. A survey of 226 Korean college students revealed that American programmes may indeed be contributing to the Westernization of traditional cultures but the impact may be more diverse than so far acknowledged. They noted "increased nationalism and anti-Americanism—which are quite contrary to the usual assumption about cultural imperialism—may have profound social, political, and economic consequences; at the same time, a decline in the absolute authority of parents, the sharing of dating expenses, and the wearing of jeans are not necessarily trivial cultural transformation" (1990: 300).

Kang *et al.* (1995) investigated the universality of value types as developed by S.H. Schwartz and W. Bilisky and evaluated the impact of television viewing on the value preferences expressed by English-proficient students in Delhi (sample size was 430 secondary/college students).

Questionnaire results indicated that value types have acceptance across diverse cultures and that institutions such as television can have an impact on value orientation. Compatibilities and incompatibilities among values within value types were found.

Cheung and Chan (1996) hypothesized that a belief in a mean and inhumane world among adolescents is generated by television's repeated projection of trivialized morality, violence, materialism, construed reality, and impulsivity. Questionnaires measuring weekly television viewing time and aspects of a 'mean world' value were administered to 402 8th to 10th grade students in Hong Kong in 1988. Results showed television's cultivation effects to be strong in the areas of trivializing morality and promoting materialism, and less strong but significant in generating a perceived reality and impulsivity. The cultivation effect is attributed to the extreme amount of violence and consumerism in Hong Kong commercial television programmes.

Dong *et al.* (1998) surveyed 439 Chinese public high school students in Beijing in 1994 and reported that US movies and television programmes were readily available and were watched at a rate of two per week. Frequency of viewing these movies and television programmes, however, did not predict acceptance of US values, though theme recognition and functionality valuation did predict such acceptance, including the value of wealth.

Kang *et al.* (1999) applied the postulates and methodology of the universal value structure theory articulated by S.H. Schwarts and W. Bilsky to mass communication, to investigate how cultural background and television viewing habits are related to viewers' perceptions of the importance of value types and to individualistic or collectivistic value structures. The data was collected through questionnaire in 1996 from a convenience sample of 362 university students in Tokyo, Japan. It revealed significant relationships between television programme genres viewed and individualist, collectivist, and mixed value types. The research concluded that television programme genre usually serves as a stronger predictor of value type importance than do demographic variables.

Varma A. (2000) based on data obtained from 11th and 12th grade students of New Delhi (78 females and 83 males, aged 15-18) using questionnaire in 1992 examined the effects of watching international television programmes on modelling; attitudes towards sex and violence; and alcohol, drug, and tobacco use. Viewers (sample size = 82) and non-viewers (sample size = 79) of international television programmes responded identically to all questions on drug use and violence and most questions on modelling, sex, and alcohol/tobacco use. However, viewers of international television expressed a stronger preference for sexual freedom, occasional drunkenness, and Western dance (rather than Indian dance) and were more fashion-conscious than non-viewers.

Willnat *et al.* (2002) investigated how young Asians and Europeans perceive the strength of United States (US) media effects on their culture. Based on Davison's third-person effect hypothesis, this research proposed that the most viewers would attribute greater effects of the US media upon others than upon themselves. The research covered 1968 Asian and European students. It included students' US media consumption, and perceptions of how exposure to the US media might influence themselves and others. Findings showed that most Asian viewers believed that their cultural values are positively influenced by exposure to the US media, while most European viewers thought that their cultural values were negatively influenced. Effects of mediated US violence, on the other hand, were perceived to be negative on self and others among all viewers. Findings also indicated that European viewers tend to believe that the US media influenced the cultural values of others more than their own, while Asian viewers generally thought that US media affect their own cultural values more than those of others.

Apart from Student' population, other studies about value orientation and television viewing supported that the values do get influenced because of television viewing. Philip Kottak (1990) in his research about Brazilian television viewers concluded, "Brazilian heavy viewers are skin colour conscious, liberal, devalue work, worry about safety, read and say that television has changed their life. Against that light viewers are more likely to consider television addictive, trust members of their personal network, consider marriage and family as important (1990: 186-187).

Stilling (1997) demonstrated correlation between English-language TV viewing and the acculturation level of Hispanic immigrants. Three-phase experiment was conducted using scale data from North Americans, Latin American immigrants to the US, Peruvians, and Mexicans. Phase-1 compared three operationalizations of Television exposure-total, content-specific, and proportional exposure—correlating these with a measure of acculturation. The proportional measure correlated most highly. Phase-2 compared three predictors of acculturation, the proportional television exposure measure and two interactions with this measure, using two intervening variables: motivation to use television for cultural information and perceived reality of English-language programming. The exposure-motivation interaction, chosen to serve in Phase-3, was the strongest predictor. Significant acculturation differences among light viewers caused by duration of residence in the US were non-significant among heavy viewers, as predicted by the electronic melting pot hypothesis.

Morgan *et al.* (1999) reviewed previous research on television portrayal of the family and used the data from the 1988 and 1994 General Social Surveys to assess relations between television viewing and judgements about illegitimacy and single parenthood, which is most directly related to the Murphy Brown controversy and is frequently touched on in many strands of political discourse. The data do not contradict Dan Quayle's contention that television viewing is contributing to fraying of traditional family values.

But there are contrary evidences in television's association with value orientation. Cashmore (1994) in the book '...and There was Television' argued that much of the concern about television's negative influences on society is without foundation. The evidence for the arguments that television promotes violence, shortened attention spans, the decline of literacy, and political indoctrination was examined and found not to pose major concerns. The meta-analysis of cross-national effects research confirms empirically the proposition that exposure to US television programming alone will not automatically generate an adoption of US values. This impact is influenced by both, research characteristics and types of dependent measure (Ware and Dupange, 1994: 955).

Ghose and Jahangirdar (1998) noted that music channels like Channel [V] and MTV have not taken generational defiance, implications of it may be borne out in surveys that have found that Indian youth are staunch supporters of 'traditional Indian family values' even as they are more career oriented, individualistic, and competitive.

The studies from five Asian countries concluded, "the portray of negative values in global or local television is related to the ownership of the channels...programme formats and subject matters contribute to the display of negative or positive values in television programmes. It is difficult, if not impossible to distinguish local from Western values, as they appear both in local and global television and they differ in degree rather than kind"(Hagiwara *et al.*, 1999: 16-17).

Cultural-Media Imperialism

The concept of cultural-media imperialism has roots in New World Information and Communication Order (NWICO); where the term was originally coined at 1973 Conference of Non-Aligned countries. Tomlinson (1991) commented, "cultural imperialism thesis is often coached in terms of military vocabulary, western culture is perceived as an invasion". He defined cultural imperialism as "use of political and economic power to exalt and spread the values and habits of a foreign culture at the expense of another culture" (Tomlinson, 1991).

Media imperialism is narrowed construct compared to cultural imperialism. Cultural imperialism includes broader social, cultural, economic and political contexts while media imperialism refers mainly to media industries. The concepts have roots in the notion that some national cultures will dominate others in a system of international exchange rigged to benefit already powerful nation-states.

Boyd-Barrett (1977) defined 'media imperialism' as a process whereby the ownership, structure, distribution or content of the media in anyone country are singly or together subject to substantial external pressures from the media interests of any other country or countries without proportionate reciprocation of influence by the country so affected. Reformulating his work Boyd-Barrett maintained that the conceptual framework of media imperialism is relevant "to find evidence of (i) considerable disparity between different areas of media activity, and (ii) considerable internal variation within any single area of media activity" (Boyd-Barrett, 1998: 170).

In 1980s disagreements with the theoretical premise of cultural-media imperialism began. The factors influencing the debate were: "media production arrangements with foreign collaborators, values used in guiding the production standards of software and the management of television organization and the "general evaluation of television programmes" (Goonasekera, 1987: 11), "actual transformation of world television system...and shifting theoretical paradigms including post-modernism, post-colonialism, and theories of the active audience" (Cunningham and Schiller, 1996), "the theory has never been applied to sufficiently comprehensive degree", commented Boyd-Barrett (1998:174) and it "needs to take account of viewers, audience preferences; audience consumption of specified cultural products as a proportion of total media and non-media consumption" (Boyd-Barrett, 1998: 168).

"In the 1990s as in the 1970s there are significant centres of media production which enjoy international influence in what used to be called developing world: for example, Bombay for Hindi films, Cairo for Arabic television, radio and print production, Mexico and Brazil for telenovelas" (Boyd-Barrett, 1998: 168) which negated premise of domination of West for media content.

Wang and Servaes (2000: 7) noted, "The assumptions of cultural and media imperialism theory of demise of local cultures and cultural industries countered, 'audience preference for local programmes, the prospering local television industries, the rise of cultural-lingual markets, and the decline of the national film industry' all point to the inadequacy of political and economic factors alone in explaining the change in the market place".

Fejes (1982) wrote that the term media imperialism "still remains vague as an analytical concept." Kivikuru and Varis (1986: 217) commented, "There does exist quite a confusion of terms and concepts." The rejection of the premise of cultural-media imperialism using varied cases led to some more confusion due to contradictory evidences (Oliveira 1986, Wang 1997, Wang *et al.*, 2000; Chandha and Kavoori, 2000). But "by the 1990s it had become evident that the theories have suffered from a lack of evidence" (Wang, 2000:7). It was also accepted that in its early days "it was too embedded in a radical political discourse against the USA" (Boyd-Barrett, 1998). Studies done in the 1980s and the 1990s showed that viewers interpret television programmes according to their social, cultural, economic and individual life experiences (Fiske, 1987; Kottak, 1980).

Fejes further remarked, "empirical social science as it has developed today is not equipped and does not have the tools to research the phenomenon of dependency or media imperialism in the manner in which these notions were originally conceived" (Fejes, 1982: 352). However, Cunningham and Schiller (1996) noted, "serious inadequacies, both in terms of theory and in terms of the reality which the theory purported to explain". Boyd-Barrett comments, "in the early days it was too much embedded in a radical political discourse against the USA" (Boyd-Barrett, 1998: 173).

Many other researchers in recent literature agreed that the premise of cultural-media imperialism requires reconsideration (Banerjee, 2002; Kavoori and Chadha, 2000; White, 2000; Roach, 1997; Wang, 1997; Sui-Nam Lee, 1995). Goonasekera (1987: 12) concluded "the influence of television on cultural values is a more complex process than what advocates of cultural imperialism make it out to be". Sinclair noted "the business of international broadcasting is driven by commercial consideration rather than for any desire for ideological domination" (Sinclair, 1997: 154).

Though cultural-media imperialism premise was negated, research studies of media text reveal the possibility of imperialism because local programmes wear global values. Richards (2000: 385) observed that in early years "Philippines television heavily depended upon programming from United States particularly because this was less costly than local production...even locally made programmes which now constitute majority of the output, bear unmistakable mark of American genres and production values".

"Today, transnational media coexist with domestic, and compete with viewers; domestic production can become even more commercial, garish and explicit than the western 'originals', but perhaps new programming formats, indigenised media products, alternative new frames, might also develop", remarked Sreberny-Mohammadi (1991: 181).

In the Indian context, Pathania (1998: 66) remarked, "Doordarshan not only carried American serials like The Valley of the Dolls in its original forms, but also Western fare which has been Indianised through cosmetic changes such as dubbing or local hosting as in the case of Disney cartoons".

Relevance of Cultural-Media Imperialism

Tomlinson (1997) citing irrefutable empirical observations and arguments like ubiquity of Western (American) media cultural goods, long history of western imperialism and centrality of capitalism as a cultural force comment that cultural imperialism thesis continues to constitute an important and widely used critical perspective today.

The metaphor of imperialism still holds strength as the fact that 'the imaginary cultural landscapes described by Appadurai are largely the product of corporate marketing practices and the repetitive consumption patterns (and patterns of desire and aspiration) that marketing encourages and supports' (Griffin, 2002). And "the operations of transnational corporate systems of technology and marketing seem to reinforce, rather than diminish, continuing inequalities between historical colonizers and the historically colonized" (Golding, 1998).

"According to numerous economic and financial reports, these levels of concentration and transnational expansion are even greater among media firms" (Herman and McChesney, 1997). And such concentration continues to extend and exacerbate gaps in communication resources and access to media and technology (including the "digital divide") among classes, regions, and nations (Golding, 1998; McChesney, 1998). Besides "for what is called newly industrializing countries cultural imperialism remains potent as an ideological defence of national governments against mass-mediated cultural influence from the West" (Sinclair, 1997).

Lately the debate has taken a new turn. Sonwalkar (2001: 506) noted, "within South Asia, the notion of Western-based media imperialism is being replaced by the India-based version, while within India, the pre-eminent position hitherto enjoyed by Hindi in the national cultural discourse until recently had given way to the suzerainty of local languages".

In the beginning of twenty-first century the debates on the cultural implications of media globalization can be generally characterised in two strands: "Cultural imperialism perspective that emphasis domination and hegemony and cultural globalization approach, which underlines cultural exchange, interconnection and diversity" (Benerjee, 2002: 518). "Language and region increasingly dominate media content and viewership, concepts of imperialism and hegemony acquire new meanings in localized settings" (Sonwalkar, 2001: 505-6).

Philo (1999) commented that localization, however, does not mean that concerns about global cultural dominance are over, given the fact that it is precisely through localization that the globalization of media, and MTV in particular is taking place. Juluri (2002) based on his research concluded that the reception of music television in India and the rise of youth culture call for a deeper understanding of cultural imperialism than a mere celebratory dismissal.

Cultural-Media Imperialism Research Studies

Cultural-media imperialism had been studied mostly from media institution perspective and television flow studies were the one to initiate the earlier theoretical formulation. UNESCO initiated early studies *Television Traffic—A one-way street?* (Nordenstreng and Varis, 1974), *Transnational Communication and Cultural Industries* (Guback and Varis, 1982), *International Flow of Television Programmes* (Varis, 1985) to examine flow of television content across the globe. The findings clearly indicated imbalance in television programme flow and showed 'one way traffic' of media content and 'dependency' of third world countries on imports of Western (mainly USA and UK) countries.

India never figured into these studies as major importer. *Variety* (1976, 1978) in its issue commented, "one of the outstanding features of India's broadcasting...[is] its determination to achieve media self-reliance...manifested in the complete absence of Hollywood television products in India, partly thanks to a prosperous indigenous film industry".

The studies on television programme flow eventually expected that the dependence had reduced because of factors like developing countries capabilities to produce locally, collaborative productions, transformation of world television system, technological advancement and so on. But the findings of the studies did not support the argument.

Straubhaar, Campbell and Cahoon (1995) looked empirically at television programmes broadcast during sample weeks from 1962, 1972, 1982 and 1991-92 for Mexico (Mexico City), Anglo and Hispanic U.S. (Chicago), English Canada (Calgary) and Quebec (Montreal). Television Programmes were categorized by genres and by where the programme originates. This research found that "US programming flows into some of the markets studied did not diminish as much from 1962 to 1992 as might have expected theoretically".

"At the same time global distribution networks emerged under the control of commercial media conglome-rates, media production (following other sectors of commercial manufacturing) became less easily identified with a few 'core' metropolises. Regional centres of media production arose and/or expanded in such places as India, Brazil, Mexico, Egypt, and Hong Kong, along with regional, 'geocultural' markets for their cultural products" (Straubhaar, 1997).

"The underlying assumption of most studies in international television flows between and among countries had been that they had cultural and economic effects in specific countries or among specific groups of viewers. Together with the concepts from media imperialism or dependency theories, this assumption had been the dominant theory that guided most data collection...it was assumed that if there is supply, there is consumption and the effect of consumption is the adoption of foreign, generally US, values and lifestyles...the consumption of transnationalised television programmes has not been included in any previously executed research of international television flows" (Servaes and Patchanee, 1998: 274-76).

UNESCO sponsored research to "include the analysis of viewer consumption with regards to country of reception, instead of researching only the sources of television flow". Five European countries, namely Bulgaria, the Netherlands, Hungary, Italy and Sweden and four Asia-Pacific countries—Australia, India, South Korea, and Philippines were included in the research. The findings of UNESCO (1994) research indicated continuation of American content in varying degree on the supply side on European and Asian television except for India. Indian television had more than 95 per cent of its programmes produced within country with little input from foreign sources (only about four per cent).

But the scenario in India changed with the advent of Cable and Satellite Television in 1990s. Chandha and Kavoori (2000) argued that national gate-keeping policies, dynamic audience preference and forces of local competition "have successfully limited the inflow of western programming in the region". By 1990s access to foreign television channels and their programmes became easier and rampant. In this context, how far the viewers get influenced or imperialised by transnational television viewing became relevant.

After UNESCO (1994) research, numbers of empirical researches were undertaken which essentially focused on consumption dimensions of transnational (at times labelled as transborder, satellite, global or international) television. Research studies in Indian context (See Melkote *et al.*, 1998, Singhal and Rogers, 2001; Juluri, 2002), Asian reference (See Goonsekera and Paul, 1998; French and Richards, 2000; Chadha and Kavoori, 2000; Page and Crawley, 2001; McDaniel, 2002) and global context using micro and macro studies (Herman and McChesney, 1997; Thussu, 1998; Hagen and Wasko, 2000; Straubhaar, 2000; Wang *et al.*, 2000; Eko, 2001; Curran and Park, 2000) were carried out.

The outcome was further confusion. Fejes's comment made in 1980s appear relevant even today, "there is very little consensus as to the basic formulation of the questions to be asked, much less agreement on methods and criteria (to research cultural dimension of media; one of the most difficult areas of communication studies)" (Fejes, 1982: 351).

Based on the research in Brazil, Kottak concluded, "Despite fears and lamentations about cultural imperialism, a cross-cultural perspective shows that television is not just a homogenizing force, instead, television is potentially a means of elaborating on the great traditions of different societies and preserving the cultural diversity that such traditions represent" (Kottak, 1990:15). Proponents of globalization support Kottak's statement but imperialism advocates disagreed with the comment.

Juluri (2002) from his research of 'MTV in India' argued that the positioning in the scholarly literature of cultural imperialism and audience studies place an *a priori* belief in audience activity. Wang and Servaes (2000: 12) observed, "in recent years the image of audience has undergone dramatic changes, from that of passive, fragmented individuals manipulated by media to that of autonomous social groups which select and interpret programme content according to their own interest and cultural background".

International Studies related to Transnational Television Viewers

"There is virtually overwhelming evidence that cultural messages are differently received and interpreted, and that meaning can be produced locally at the point of reception... [but] exact impact of television on communities is shown in the ambiguous nature of research results" (Robertson, 1994: 46). Way back in 1984 Katz and Liebes in their research about cross-cultural reception of American serial *Dallas* noted, "during and after a [television] programme, people discuss what they have seen, and come to a collective understanding. It is via such a understandings, we believe, that the messages of media enter the culture" (1994: 419).

With the wave of globalisation and dramatic restructuring of national media systems in many countries lot of research studies were published about influence of transnational television on the cultures (Herman and McChesney, 1997; Melkote *et al.*, 1998; Goonsekera and Lee, 1998; Thussu, 1998; French and Richards, 2000; Curran and Park, 2000; Wang *et al.*, 2000; Page and Crawley, 2001; Singhal and Rogers, 2001; McDaniel, 2002).

Kottak (1990) in his research dealt with Brazil and United States of America covering cultivation effect (influence of content on heavy viewers) and the behavioural effects (social repercussions in the Brazilian communities) of the television. He dealt with both content and viewing circumstances affecting Brazilian society. He concluded that "the television impacts in three stages. Stage one is initial one of strangeness and novelty when gazes are usually glued to the set. The medium rather than the message is mesmerizing. In second stage people become more accustomed to and comfortable with television. They begin a process of selective acceptance and rejection, interpretation, and reworking of television messages. In third stage community saturation and length of television exposure increases, clear statistical correlation between tele-viewing and other variables diminishes". He also reported negative attitudes towards television increased with higher income and with years of exposure (Kottak, 1990: 140).

Elasmar and Hunter (1992) in a meta-analysis involving thirty-six pieces of transnational broadcasting research from 24 countries, found that "transnational broadcasting only had a limited impact on viewers. There was no clear relationship at all between exposure to foreign programmes and viewers knowledge, ideas, attitudes, values, or behaviour, still less a causal relationship".

Ware and Dupagne (1994) conducted meta-analysis of studies on 'effects of US television programmes on foreign viewers. Out of the thirty-four studies collected seventeen were used for the final analysis as remaining did not "meet the statistical requirements of meta-analysis". They concluded that "US imports have a small but statistically significant impact on foreign viewers and measures dealing with preference for American goods and attitudes towards America displayed the largest effect size". They also remarked, "when taking research characteristics into consideration...studies using English questionnaire displayed larger effects".

Earlier cited UNESCO (1994) research concluded that "the actual viewership of transnational programmes was very small and it differed from country to country". Servaes and Malikhao (1998: 290) discussing the UNESCO research concluded, "the climate in which television consumption takes place, and other factors such as scheduling of programmes, the culture, the socio-economic development may be stronger determinants of transnationalisation of television consumption, than demographic criteria such as gender, age, education or urbanization".

French and Richard (1996) discussed about countries without the capabilities to produce and distribute TV programmes. It is argued that "the need for many African and Asian nations to depend on Western TV sources is generating tensions between commercial realities and the desire to preserve cultural identities".

Straubhaar, Campbell and Cahoon (1995) studied five cultures (Mexico, Hispanic U.S., Anglophone U.S., Anglophone Canada, and Francophone Canada). The findings showed that viewers continued to be interested in the cultural production of the US, particularly when they live next door to America. It also supported the idea that viewers are very interested in cultural proximity within both cultural-linguistic areas (such as Anglophone or Latin American cultural markets) and within sub-national markets defined by language and culture (such as Quebec). The programme genre analysis showed that the cultural proximity as a strong motive of viewing certain genres of television. These particularly include news, public affairs, and melodrama or soap opera and, to lesser degree, music.

The cross-cultural research (Goonasekera and Lee, 1998) conducted in five countries (India, Malaysia, Philippines, Hong Kong and Japan) also reported (Hagiwara *et al.*, 1999) elsewhere focused on "the flow of global television in terms of actual exposure and audience perception of its impact". The methodology included general historical survey of the country's television industry and its regulatory issues, a questionnaire survey of around 200 viewers and a content analysis of global as well as local programmes. The findings of audience survey indicated that most Asian viewers consider global television productions to be better than local ones for documentaries, cartoons and sports. For news and information programmes, sitcoms and films the judgements varies with country. Indian viewers preferred local films; Malaysian viewers enjoy sitcoms from the West while Filipino viewers consider their own news and information programmes to be better. Most of the Asian viewers consider global television more of good thing than a harmful influence.

Hargreaves and Mahdjoub (1997) examined television viewing habits of Muslim families with satellite dishes. Based on a qualitative survey conducted among 115 viewers in 30 households, mainly of Maghrebi origin, living in Montbelard, France. Viewers were asked what they had expected of satellite programmes, how their viewing habits had changed with the installation of satellite television receivers, and how this had affected their personal and family attitudes. Findings revealed major differences between first- and second-generation Maghrebi origin viewers, as well as significant variations between genders.

Lee (1997) proposed three-tier model of consumption for television in non-English speaking countries where local programmes are weak. The first tier of programmes consumed by audience was "local programmes". The second tier consisted of "culturally proximate programmes with popular genres like those in first tier". The third tier comprised "programmes with less culturally proximity, and the genres are usually information-related, international sports to productions with high technical skills such as science programmes and animation". Singhal *et al.* (1998: 252) remarked, "the active preferential choice made by an individual in the audience to view international, national or regional television programmes, a choice that typically favours the latter two when available because of their greater cultural relevance for the audience member".

Lee (2000:188) concluded, "the general findings of nearly all empirical studies point to the fact that viewing global television has little impact on viewers, not to say changing their deeply entrenched values. It was found that global television was mostly shown in fringe hours in non-English speaking world and viewers usually preferred local to foreign programmes". Lee's point may not hold true in the era of round the clock satellite television channels broadcasting available in the most of the Afro-Asian and South American countries. Further the viewers, "have different motivations—to get information unavailable at home, to get better entertainment, or to catch up with topics related to some popular foreign programmes, movies or celebrities" (Lee, 2000: 188) and "the audience is always in search of options and alternatives and has no particular resistance or inhibition to programmes originating from an alien source provided the programmes is interesting and attractive and not repulsive" (Bhatt, 1994: 20).

Transnational Television Viewers Studies in India: Transnational television is India is about a decade and a half years old. The studies from Anthropology (Johnson 2000), Psychology (Varma 2000), Sociology (Chauhan 2001, Datta and Alwe 1999, Rao and Raghavan 1996), Women's studies (Mankekar 2000) Cultural studies (Juluri 2002, Butcher 2001,

McMillin 2001, Monteiro and Jayasankar 2000, Rajagopal 2000, Mankekar 1999, Mitra 1993) International Communication (Thussu 2002, Thussu 1998), Media Studies (Malhotra and Rogers 2000, Muppiddi 1999) have dealt with transnational television. Few Social Scientists (Sinclair 1998, Dube 1990, Joshi 1998, Goonasekera and Lee 1998 and Melkote *et al.* 1998) also studied influence transnational television on Indian culture. Few Indian researchers through their foreign university affiliations (Mitra 1993, Muppiddi 1999, Mankekar 1999, McMillin 2001) also carried out studies.

Transnational television consumption studies in India can be classified into following three categories:

(i) Studies related to 'identity' and nationhood
(ii) Studies to assess viewership or access
(iii) Studies profiling viewers and their concerns

(i) Studies related to 'identity' and nationhood have been mainly conducted from cultural studies perspective using surveys and ethnographies as methodology. They are conceptual debates (Shield and Muppiddi, 1996; Fernandes, 2000) or audience reception research (Rajagopal, 2000; Gupta, 2000; 1998, McMillin, 2001; Butcher, 2001; Juluri, 2002).

(ii) Studies to assess viewership or access had been carried out to estimate reach, coverage, and preferences of viewers because of commercial interests of private satellite television channels. Studies conducted in this category were mainly for advertisers and to some extent for programme producers. Television Audience Measurement (TAM) people-meters, Doordarshan's Audience diary, studies by market research firms are some of the examples of these types of studies (Bhatt, 1994; Madhvani, 1998; Businessline, 2002). Many of these studies are not available for public distribution or reference due to commercial considerations.

(iii) Studies profiling viewers and their concerns: Most of the studies profiling viewers and their concerns belonged to 'sociology, psychology and anthropology' (Agrawal, 1986) since they were the initial disciplines to initiate communication and media research in India. The trend continued with the research studies related to transnational viewers.

Rao and Raghavan (1996) reported a countrywide social audit of functioning of mass media in their research conducted between October 1994 and February 1995 covered 2464 TV homes from seven states and found cable and satellite television in 40 per cent households. Questionnaire, focus group discussion and convened colloquial were used to collect data. The research reported that television claimed highest share of viewers compared to other media. 'Entertainment' was the reason for opting for cable connection by 67.31 per cent of the cable households, followed by for watching films, for news awareness and English movies and serials. People not opting for cable gave reasons like enough entertainment on Doordarshan, children's studies, and non-availability of cable operator in area. The research also reported that four per cent of the viewers had disconnected cable temporarily or permanently of which 67.1 per cent had done it because of children's studies. Cable and satellite television households reported to be spending 246 minutes compared to 210 minutes of Doordarshan households.

Shah (1997) used descriptive technique to narrate hype and hypocrisy of television in urban India. She described developments of video and cable television in India and its implications on the Indian society.

Rao and Melkote (1998) reported a research of cable subscribers and non-subscribers in Bangalore city. The research covered 600 households. Data was collected using a cross-sectional survey method with the help of a questionnaire. Significant differences between cable (and satellite television households) and non-cable subscribers were found in terms of income and frequency of viewing all programme types except Western serials. The level of viewing satisfaction differed mainly in music and Indian serials category between the two groups of viewers.

Joshi (1998) conducted a survey in February 1994 covering 263 viewers in Mumbai (Bombay) and Chennai (Madras). He also content analysed the programmes of Doordarshan, Zee and STAR Plus as a part of larger project reported earlier (Goonasekera and Lee, 1998). The content analysis did show satellite channels showing values objectionable to Indian way of life while Doordarshan telecast the most purposeful, socially relevant and meaningful programmes. Joshi (1998: 37) concluded, "there are, without doubt, some genuine worries and concerns about the entry of foreign satellite programmes which are widely accepted and welcomed with full confidence that Indian society can meet the software and hardware challenges and also the Indian value systems, and culture are a strong enough to withstand any outside undesirable influences" (Joshi, 1998).

Gupta (1998) studied impact of television from semiotics and cultural perspective with a view to examine "cultural alienation and distortions caused to Indian culture due to ill effects of television". She conducted comparative survey of 180 households with and without cable (and satellite television) connection. In addition she interviewed 50 cable operators from seven localities in West Bengal using questionnaire. Gupta (1998) reported that C&S TV viewers watched 15.2 per cent more television compared to non-C&S TV viewers. She concluded "while there is a sense of guilt associated with the act of watching television, the viewers seems to be using television programmes to redefine their lifestyles in various ways" (1998: 136).

Datta and Alwe (1999) conducted a sociological research to examine impact of satellite television on family interaction patterns with children between 10 to 12 years of age and their parents in Mumbai. 60 children from nuclear and joint families with and without (cable and) satellite television were randomly selected. Interview guide for children, questionnaire for parents and focused group discussion with all of them was conducted to collect data. The findings indicated that the families not having (cable and) satellite television do feel segregated and isolated and feel that they cannot avoid for long not taking cable connection. Though no definite statistical relationship could be established between television viewing and interaction patterns, Datta and Alwe (1999) observed "a developing trend of reduced quality interaction is beginning to appear".

Mankekar (1999) carried out ethnographic research in New Delhi *bastis*. The research was guided by "feminist anthropology and ethnography and feminist interventions in cultural studies of mass media". She recorded responses of 'upwardly mobile, middle-class urban women'. Mankekar dissected both the perceptions as well as reactions to television messages. The descriptive account supports "active audience" premise and confirms the role of television in helping the viewers "imaginatively connecting with larger connectivities such as community and nation".

Muppiddi (1999) studied audience preferences for a regional language channel over national channel. Within uses and gratification theory, a survey of 400 households using questionnaire in Hanamkonda city of Andhra Pradesh was carried out. Two television channels—Doordarshan and a private Telugu channel Eenadu—were selected. Muppiddi concluded, "while viewers do prefer Eenadu television, this preference could have arisen from a preference for the language, quality of programme content, availability of preferred programming types or through identification with the socio-cultural context...more research needs to be done before all factors that influence such viewing behaviour can be elaborated and before reasons for preference for a particular channels can be completely explained".

Johnson (2000) examined television and social change in rural India by researching two mountain villages in Western Maharashtra (Danawli and Raj Puri) with reference to influence of television on various social, economic, and political processes. Data from ethnographic fieldwork describes the unique characteristics of television that makes it an important agent of cultural change. His discussion included consumerism, urban modelling, restructuring of human relationships, demographic change, linguistic hegemony, migration, and the emergence of an information underclass. Social change is examined at both the structural and psychological levels. Johnson argued that the village audience is an active and vibrant participant in the use of media.

Doordarshan Audience Research Unit of Jalandhar conducted a field survey in 2002 of randomly selected 800 cable (and satellite) TV households (in proportion of 49% males and 51 % females) in Amritsar, Jalandhar, Patiyala and Ludhiyna, four major cities of Punjab. The findings revealed that combined channel viewership of DD Punjabi channel was most popular among the audience (84%) followed by STAR (75%) and DD-1 (84%). The private language channels trailed behind national language channel as against 84 per cent preference for DD Punjabi 32 per cent reported preference for Lashkara, private channel. The other findings indicated high preference for movies, feature films and telefilms (53%). (http://www.thehoot.org)

Butcher (2001) explored the impact of transnational television on youth of India from cultural studies perspective. Examining Indianness and identity she concluded, "the argument surrounding the effects of transnational television are made complex by examples of social conceptualisation of viewing and there is compounded substantial data that suggests that other technologies are also influencing the readings and subsequent understandings of transnational television programming in India".

Juluri (2002) conducted a reception research of nationalistic Indian music videos with 42 participants between May to August 1997. The findings suggested that transgenerational and pronational youth culture emerging in India was not evidence of audience resistance but, on the contrary, the form in which global hegemony is indeed taking place.

Verma and Larson (2002) examined effects of TV on youth, taking 100 urban, middle-class Indian families to elucidate the 'context of use' of TV. Mothers, fathers, and 8th graders carried alarm watches for one week and provided 13,674 reports on their activities and subjective states at random times across waking hours when signaled. It was found that TV viewing for youth is typically a family activity, occurring in a context in which 'parents' supervision and influence is likely. Adolescents' rates of viewing were correlated with parents' rates of viewing, and fathers' type of employment. TV viewing of middle-class Indian youth is typically a relaxed antidote to the stresses of the day that they share with their families.

Scrase (2002) examines if globalized television culture has led to a transformation of middle-class Bengali cultural identities. Ethnographic fieldwork carried out during 1998-2000 explored specific themes: television and rampant consumerism; the impact of 'Bollywood' film culture; television and the portrayal of women; and the way satellite and cable television programmes, especially popular cultural programmes (e.g. music videos) and advertisements, impact on established cultural patterns and activities. The research showed that while a public, more liberal tolerant culture prevails in Bengal, the private world of the family nevertheless retains many aspects of traditional moral and hierarchical principles. Divisions of opinion about the social impact of television among the middle classes are explained in terms of the nature of middle-class cultural formation in Bengal since the time of British colonization of India.

Dhruva (2004) carried out a survey of 100 married women in the age group of 25 to 35 years of age residing in urban slums of Mumbai to examine influence of television serials in their lives. The findings indicated that television serials influenced daily routines and personal behaviour of women where as they negated any influence on family relationships. The study also states that in spite of the lower economic backgrounds the respondents reported to like the 'rich lifestyles' projected in the television serials.

In Brief

Review of secondary literature reveals clear association between television and cultural values. Cultural imperialism and cultural globalisation are two ends of media in global society debate. Cultural-Media imperialism includes cultural as well as media influences of some national cultures dominating others in a system of international exchange benefiting already powerful nation-states. Cultural globalisation views media as an integrative force rather than divisive one.

Since its inception in 1970s cultural-media imperialism concept, faced criticism by 1980s for variety of reasons. But even by the beginning of the twenty first century it still remains relevant concept because analysis of television text suggest local productions wearing global values and global media propagating capitalist consumerism, alien to Eastern cultural landscape. The review of research studies reconfirm confusion and contradictions of transnational television consumption research.

3

Television in India

Television in India is about to celebrate its golden jubilee. It is a dramatic story of twists and turns where private party entertainment approach overtook the public sector 'education' centred initiatives and the race reinvents itself by twenty first century when both coexist having unique identity of their own. Today India has the largest public service broadcasting network in the world, Doordarshan, besides educational television variations in form of EDUSAT project, Gyanwani, University Grants Commission's Countrywide Classroom and School Broadcast, all under State manage its presence along with 300+ local, national and multinational satellite channels.

History of television in India dates back to 1959. The first telecast originated from a makeshift studio at *Akashvani Bhawan* (Broadcasting House), New Delhi on September 15, 1959. UNESCO gave a grant to start an experimental telecast in and around Delhi (UNESCO, 1953, 1964; Bhatt, 1994). It aimed to provide "social education" initially through one hour transmission twice a week. It started in form of a (educational broadcast) project with 250 television sets installed in 144 schools (in and around Delhi) having an electrical supply. It was proposed to cover all the secondary schools in Delhi in a due course. By the end of 1962, 388 television sets had been installed in 192 schools. Fifteen thousand students viewed science lessons and about 50,000 students took advantage of language lessons (GOI, 1963:130). Regular one-hour television transmission service started after six years in 1965, which included a daily news bulletin. Little happened on Indian television scene in terms of expansion prior to Satellite

Instructional Television Experiment (SITE), which marked the beginning of satellite television in India.

Beginning of Satellite Television in India

Satellite Instructional Television Experiment (SITE), was the first ever techno-social experiment of direct satellite broadcasting in the world. It was initiated on August 1, 1975 for a period of one year. Doordarshan (1999) reports that it was incidentally the first attempt anywhere in the world of using satellite technology for social education and national development. "From all accounts SITE was a successful experiment and it achieved both its social and technical objectives thereby paving the way for a satellite-based television system in India", DECU (2000:18).

1980s witnessed operationalisation of satellite television in India after the launch of Indian National Satellite (INSAT) in 1981. Satellite broadcasting for national and regional centres became regular feature of Doordarshan with INSAT. Satellite television in that time period meant television transmission using satellite. Early 1990s witnessed dramatic changes as satellite broadcasting which was 'national' prior to 1990s became transnational post-1990s. The nomenclature of 'satellite television' also became 'Cable and Satellite Television' (C&S TV) as cable became delivery mechanism for private satellite television channels.

Television Ownership

There were only 60 television sets in the country in 1962 (UNESCO, 1964). The number increased to 2,75,424 in 1974 and within two years almost doubled to 4,76,026. The last official figure of television sets in the country in 1984 was 36,32,328. No official figures are available now since the policy of licensing of television sets was withdrawn in 1985. Table 3.1 shows the growth of television households in the country. It also indicates difference in the estimates as NRS 2006 reported 112 million TV sets in India compared to 103.3 million as per IRS 2008.

In 1994, Senstrup and Goonasekera (1994) reported that Korea and Australia where 99.9 and 99 per cent of the households owned television sets, wherein in India only 20 per cent households owned television. By 2002, 42.3 per cent of Indian households had television sets but three-fourth (76.1 per cent) of them were in urban India (Doordarshan, 2002: 33).

Table 3.1: Growth of Television Households in India

Year	*1985*	*1990*	*1995*	*1997*	*1998*	*1999*	*2002*	*2005*	*2006*	*2008*
No. of TV Sets (in millions)	6.8	27.8	52.3	57.7	69.1	75.5	79.4	108	112	103.3

Source: Doordarshan 1997/2002, NRS 2005/2006, IRS 2008.

Beginning of Video and Cable Television

Video boom took place in India in 1980s. "In 1982, along with television, the Indian government had liberalized imports of video cassette recorders by lowering duties and allowing travellers to bring them as a part of their personal baggage...By 1987, there were an estimated 10,000 video parlours, 12,000 video coaches and 11,000 hotels with VCRs in India" (Shah, 1997:56). "In less than five years, from 1984 to 1989, the penetration of video moved from the capital down to the districts and villages" (Agrawal, 1991).

Like video boom, cable television in India dates back to early 1980s. Enterprising individuals in apartment blocks placed a video in their homes or in their garages and started offering a cable TV service to the viewers in their building or vicinity. The fare available was Hindi and English movies and pirated western comedies, music and game shows. Video viewing in place of group activity, became community activity at a monthly payment. Initially cable TV was restricted to the upper crust of society in the major metros and towns. Shah (1997: 58) notes, "In Bombay [now renamed as Mumbai] a couple of enterprising young men led the way by using homemade splitters and amplifiers to connect several buildings in upper class residential areas...to a few strategically placed VCRs (Video Cassette Recorders) on which they showed films and foreign serials...by 1989 there were an estimated 1000 cable operators in Bombay". Mumbai with its dense population and chawl housing, offered ample scope for growth. Developments of cable television in Mumbai have been reported by many researchers (Agrawal 1991, Bhatt 1994, Shah 1997).

Growth of cable television in urban India indicated a rise from 0.41 million in 1992 to 68 million in 2006 as shown in Table 3.2. Doordarshan (2002) reported four per cent growth in television household and 14 per cent rise in C&S TV homes in urban India for the year 1999. Today, India has more cable connections than telephone lines.

Table 3.2: (Estimated) Growth of Cable and Satellite Television in India

Month/Year	*No. of Household (in millions)*
1/1992	0.41
2/1993	3.30
10/1993	7.23
6/1995	9.30
12/1996	11.0
11/1998	18.5
1999	29 (NRS) 25 (IRS)
2002	37.9 (NRS)
2003	40
2005	61
2006	68
2008	66 (IRS)

Source: Doordarshan 1997, *The Economic Times* 1998, IRS* and NRS** Figures* IRS—Indian Readership Survey; **NRS-National Readership Surveys.

New Face of Satellite Television

Until 1991, the television broadcasting in India meant the State controlled nationwide Doordarshan network. After 1991, with the availability of telecast of Gulf war by CNN, satellite television became 'transnational television'. CNN noted major boost in its viewership from 11 million viewers in US alone before war to 75 million households in over 100 countries (Doyle, 1992:80). The cable operators through Asiasat-I could access the STAR channels launched in February 1992. Zee channel introduced in October 1992 became the prime mover of the expansion of C&S TV in India.

Thomas (1998) commented that the expansion of the Indian television "stem from two historical movements (both having close linkage with satellite television) one, period coterminus with the SITE (1975-77)...and second the era of media liberalization which started in mid-eighties and accelerated in the early nineties with the entry of cross-border state broadcasting...the unfolding processes that continue to affect the policies, structures, channels, ownership patterns, programming and audience of broadcasting" (Thomas, 1998: 20).

"Satellite based broadcasting, coupled with cable narrow casting has demonstrated that government inaction is the best condition for private enterprises to flourish, and they are now unstoppable" (Reddi, 1996: 243). According to *India Today* dated March 31, 1992 an internal report of Ministry of Information and Broadcasting had predicted as early as February 1991 that "programmes specifically targeted at Indian viewers are likely to be beamed from foreign satellites in the near future". Reaction of Government of India was "there is no threat" (*Indian Express* 1992, *The Times of India* 1992).

Government Reaction to C&S TV

The appointment of the "Vardhan Committee" in 1991 was the first step by the then Congress Government to respond to the new developments of private satellite television. The Committee re-examined the Prasar Bharati Act, 1990 to tackle the issue. Later the United Front Government in 1996 set up "Ram Vilas Paswan Committee" to decide National Media Policy. Ram Vilas Paswan Committee submitted 104-page working paper with 46 recommendations encompassing public and private electronic media, newspapers, news agencies and film. Some of the recommendations were incorporated in the Broadcasting Bill introduced in Parliament during May 1997.

The Broadcasting Bill made it mandatory for all the television channels to uplink their programmes from the Indian Territory and banned cross-media ownership. But in March 1999, all Indian broadcasting companies were permitted to uplink through their own earth stations. This policy was further liberalised in July 2000 to allow foreign-owned companies also except for foreign news-channels.

Government of India in the Ministry of Information and Broadcasting, by notification *vide* number S.O. 39(E) dated the 14th January, 2003, envisages "transmission of programmes through an addressable system" hereinafter referred to as the Conditional Access System (CAS) made "it obligatory for the cable operators in the whole of Chennai and Kolkata Metropolitan areas and the areas covered by the Municipal Council of Greater Mumbai and the National Capital Region of Delhi to transmit or re-transmit programmes of every pay channel through an addressable system within a period of six months from 15th January 2003" . The deadline got extended twice and finally, in March 2006 Delhi High Court has given a ruling to the effect that Conditional Access System (CAS) is mandatory in Southern Part of Mumbai, Delhi and Calcutta for receiving pay TV channels.

McDowell (1997: 168) remarked that more channels, cable television distribution regulation, programming changes highlight Government of India's response and policy choices in 1990s.

From Satellite Television to Transnational Television

SITE was the first step in the direction of satellite television in India. Indian National Satellite (INSAT) system in 1981 boosted the expansion of television in India leading to introduction of colour television in India in 1982. India established hybrid TV broadcasting system over the years, which is supported by satellites as well as by terrestrial transmitters. Satellite television entered in the Indian households with a new face in 1991 in the form of Gulf war telecast by CNN. The cable operators through Asiasat-I could access the STAR channels launched in February 1992. Zee channel introduced in October 1992 became the prime mover of the expansion of Cable and Satellite television in India. Growth of transnational television in form of such private channels has given new dimension to the personality of Indian television. It was different from the point of view of ownership, programming content, objectives for telecast and the whole of the other associated variables. The new form of satellite television is largely privately owned and due to its commercial agenda has 'market first' approach. The Indian viewers suddenly had spurt of variety of satellite television content including foreign content in appreciable quantity.

Transnational television in the form of foreign programmes began appearing on Indian television in the late 1970s. 1980s witnessed British, United States or German television programmes. In 1990 transnational television in form of Cable and Satellite Television (C&S TV) had the nature and the scale of foreign programming on Indian television beyond comparison. Beginning of twenty-first century was remarkable as many foreign transnational television companies started creating 'localized' content exclusively for Indian television audiences. Lot more of licensing, copycatting is being observed in this time of history. The shift post-1991 telecast of Gulf war by CNN was remarkable as far as Indian television viewership was concerned.

Many surveys indicated a shift of viewership from national channels to transnational broadcasts. "Among the first subscribers to satellite television in 1991 were the Westernised upper-class households" (Gupta,1998:67)...and "the dissatisfied elite of nations with government controlled electronic media" (Kishore, 1994: 97). Private satellite telecast in India coincided with the entry of multinational companies. Presently the Indian Cable and Satellite television viewer has access more than 300 channels. Now Satellite television in India is transnational in nature unlike its initial years. Doordarshan (2002) noted that 42.3 per cent of the total households in the country having television sets, 47.8 per cent of them had cable connections. So, today transnational television is all pervasive and omnipresent on the Indian television scenario. The figures in 2005 too indicate "a gap in reach and room for growth. There are 150 million households, but only around 80 million own television sets. In that only about 45 million households have cable connection" (Frames, 2005). As has been observed by the TAM (Television Audience Measurement) study in February 2007, upper Socio-economic Class viewers are the first one to adopt to CAS and purchase Set Top Boxes to view transnational television. So, it is history repeating itself.

Present Scenario of Cable and Satellite Television

The review showed that figures of ownership of television as well as cable and satellite television sets were estimates and at times contradictory. After Cable TV Network Regulation Act 1995, the number of cable operators who registered themselves with the post office was about 16,000 in the country. However, there were estimates of about 30,000 such operators. The cable industry is highly unorganized and heterogeneous.

"National Readership Survey 1997 data showed that 47 per cent of Maharashtra homes had television sets. As far as Mumbai was concerned 83 per cent of the households had television sets and 61 per cent had cable connection" (*Cable Waves*, 1998: 30). It is believed that over 85 per cent of the television households in Mumbai have cable TV connections today. The population density of Mumbai city also allowed for faster cable penetration. Cable and Satellite television has reached slums as well as high rises and C&S TV household in a city pays much less per month (Rs. 125) than the national average (Rs.157). Mumbai city having implemented CAS since December 2006 might have undergone some changes in subscription. According to one of the reports of indiantelevision.com in February 2007, there are 7.96 million cable homes across the three metros with 1.63 million (approximately 21 per cent) falling under the CAS-mandated zones. Mumbai has 3.25 million cable homes while in Delhi it is 2.61 million and in Kolkata 2.1 million. Mumbai apparently have highest CAS adaptation with 1,39,000 subscribers buying STBs on a CAS home of 5,48,000.

Present delivery mechanism for Cable and Satellite Television includes MSO (Multi Service Operators) who receive signals at the control room (by estimates Mumbai alone has 45 control rooms), which in turn are redistributed by area distributors/franchises to the households (Mumbai has 2500 distributors and four MSOs). Entertainment branch of Collectorate provided figures of cable points in Mumbai city, which amounted to 1,37,641 in 2001. The department officials remarked that it was underreported data. Varma on the similar lines commented, "one survey by Maharashtra government between January and March 2000 Mumbai alone has approximately three lakh cable homes...Cable operators scoffed at the government survey and admitted that there are more than ten lakh connections in metropolis" (Varma, 2000: 14) as also discussed in Sampling Section (p. 67).

Transformation of Transnational Television

Language and culture has become crucial market forces for transnational television across the globe. Even in India, majority of the private satellite television channels were forced to 'Indianize' the programming content or expand in form of regional language channels to capture the market. Impact of private satellite television channels in local languages in the Southern part of India has been notable and well documented (Rao and Melkote, 1998; Muppiddi, 1999; Juluri, 2002).

In 1997 there were no Indian language television channels except *Doordarshan* and Zee TV. In 1999 the multinational stakeholders like STAR and Sony decided to go Hindi to compete with Zee TV. In 2000, exclusive Hindi channels like SAB and Sahara were also launched having Indian ownership. After Zee News in 1997, year 2003 saw launching of many news channels in Indian languages and 2004 and 2005 witnessed business news channels. Of late there has been exclusive channels for religious discourses, health, real estate, teleshopping, travel, food and so on available to Indian audiences. Today cable industry regulations, hike in paid subscriptions, rise and fall of game shows, popularity of value-loaded 'family dramas' in the form of K-serials and debates about Conditional Access System has become history. The television viewers in India like other parts of the world has been segmented. The nature of private satellite channels has varied from music, news and current affairs, information, movie, religious discourse, and regional language, 'general entertainment' and educational-infortainment. The data reveals that there are 18 satellites beaming 300+ television channels of which about 40 per cent are English language channels.

Today an Indian C&S TV viewer is exposed to four types of programming: International, National, Regional and Local. Most of the international programming is in English and dubbed in Hindi or English. National programming includes: Public Service Broadcasting (Doordarshan—DD1 and DD2 channels) and a whole host of private satellite television channels. Regional programming covers official language channels of Doordarshan for States and private satellite television channels in most of those languages. Local programming content is through cable network in the form of live telecast of events, and daily local news bulletins (covering local events like fairs, religious discourses, civic elections, community games such as Bingo and favourite local sports), music on demand/dial in juke box, films and local advertising. Practically, every cable network has at least two cable channels—sometimes both of which screen pirated feature films and film songs in various languages including English. There are now large cable TV networks having channels like: Siti Cinema (Siticable), CVO (inCable), C-News (Hathway) in Mumbai.

4

Methods and Techniques

Television research in India has attracted various academic disciplines of social sciences. Television has been studied by sociology, psychology, anthropology, economics, market research scholars using varied quantitative and qualitative methods.

Present research followed an exploratory research design as it was thought to be most suitable to assess the influence of transnational television. Since limited empirical researches exist about influence of transnational television among the multilingual Indian society exploratory design also provided broader scope and larger canvas for mapping the influences.

Survey method was considered to be most suitable in order to have an in-depth understanding of the possible television influences and to gather overall context of television viewing, family "group interviews" were conducted which helped to gain better insights into the consumption of television by the entire household.

Uses and gratifications approach was used in the research. It assumes that the viewers are active; freely select the media that best satisfy their needs. Further, media are not the only source of gratification for making value judgements on various issues of any significance. This book explores the relationships between patterns of gratifications and influence of transnational television on values of individualism and consumerism of the viewers.

Locale of the Research: Journey of Bombay to Mumbai

Mumbai is Asia's 13th largest and India's largest megapolis. It being multilingual, metropolitan environment is a commercial capital located on the southwest of the western region having largest number of Cable and Satellite television (C&S TV) households. "Greater Mumbai population is so varied and so representative of practically all States of Indian Union and all religion and social groups that it is rightly called epitome of India" (Chaudhary, 1986:735).

The City of Bombay was founded in 1661 when its population was around 10,000. By 1872 when the first decennial Census was taken, the population of the City was 651,632. "The growth of Bombay was not uniform during the last quarter of the nineteenth century due to periodic famines in the hinterland compelling the rural inhabitants to come to the City and periodic epidemics like plague, forcing the City dwellers to seek shelter in the rural areas in large number...By 1951 the population of Greater Bombay increased mainly due to the influx of refugees following the partition of the country and also due to the merger of a number of towns with Greater Bombay, increase in trade and spurt of industry in a big way" (Census, 1991:43).

The population censuses from 1951 to 1981 for the Greater Bombay Municipal Corporation (GBMC) limited itself to GBMC covering Mumbai City, Mumbai suburbs and extended suburbs. In 1991 Census Greater Bombay Urban Agglomeration (GBUA) was formed which was further expanded in 2001 Census. The population of Mumbai as on March 1, 2001 was 1,19,14,398. In the year 2000, State Government renamed 'Bombay' as "Mumbai".

As per the estimates and projections about the largest urban agglomerations of the world, with the population of ten million and above, prepared by the United Nations in 1990, Mumbai stands sixth. "Although it has the strongest and most diversified economy of any of the India's metropolitan cities and higher level of income and a lower incidence of poverty with very high literacy rate, Mumbai continues to have serious infrastructural deficits" (Census, 2001: 61).

Prior to 1961 the erstwhile State of Bombay predominantly belonged to Gujaratis and Marathis. On the linguistic basis, it was divided and Gujarat and Maharashtra came into existence in 1961. Since Maharashtra had a predominantly Marathi speaking population, Gujaratis residing in Mumbai—some for several generations became migrants. However, together Gujarati and Marathi speakers constituted more than half (55.6 per cent) of the population of the Mumbai District. Available linguistic data as per 1991 Census (Census 2001 was not available) indicated 41.7 and 13.9 per cent persons were Marathi and Gujarati speakers respectively in Mumbai District.

Sampling

Four factors, namely language spoken in the household, ownership of C&S TV, adult member in the age group of 20 to 50 years and resident of Mumbai City were taken into consideration while selecting viewers for the research. Mumbai City population of 3,326,837 persons (Census, 2001) having 53.4 per cent in the age group of 15 to 45 years was taken into consideration for the present research.

The decision about sample size faced difficulty of information about three dimensions: age data, linguistic data and C&S TV household data in Mumbai City. Visits to Mumbai Municipal Corporation, Entertainment and Election branches of the Collectorate office, Census Office as well as offices of few Cable Operators did not help. Also data about linguistic groups for each Ward were not available. Only the Voter's name-lists were available with the Election Office and in Ward offices. The data about C&S TV households in Mumbai City was misleading as Collectorate office entertainment branch personnel (who maintain records of C&S TV connections in the City) orally reported that the cable operators for taxation and other purposes under-report the number of C&S TV connections. Abraham (1997) also made a similar observation, "Entertainment branch of the Collector Office officials themselves admit that at least five times this number of houses must be watching various satellite-linked channels through [cable] operators". Sample size was to be decided for the present research and as noted earlier no specific data were available about the universe of the research.

It was decided to draw 200 sample units each for both the linguistic groups making the total sample size of 400 households. The size for the family group interviews was decided considering the constraints of distances, time, money and resources. Two families per administrative ward for each linguistic community were purposively selected for better ward representation and one household of inter-community marriage was also identified. In this way total 17 households were covered for family group interviews.

Multi-stage sampling procedure was employed for the selection of the viewers, whereas "family group" was selected purposively to maximize representation of hitherto diverse households of Mumbai City. **Visual 1** shows coverage of the research on the Map. Of the total nine administrative wards of Mumbai City, five administrative wards were covered; three administrative Wards having both the linguistic communities and two wards with one community each as shown in Table 4.1. For the selection of individual viewers six stage random sampling procedure was followed as shown in Fig. 4.1.

Fig. 4.1: Multi-Stage Sampling Procedure

1st Stage	: Administrative Ward (Lottery Method)
2nd Stage	: Election Ward (Lottery Method)
3rd Stage	: Geographic Location for first unit (Purposively)
4th Stage	: Identifying Housing Society/Chawl/Flats/Slums (Random)
5th Stage	: Locating household having sample characteristics (Purposive)
6th Stage	: Male and female equal ratio through quota (Snow Ball)

VISUAL-1

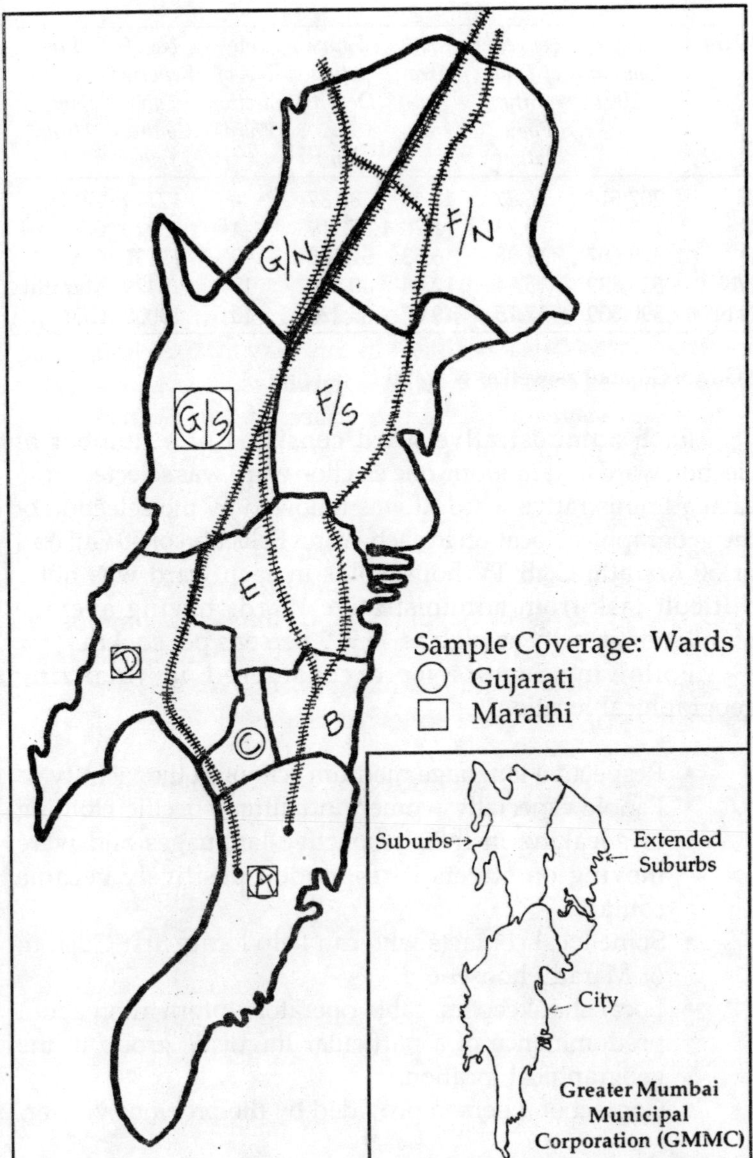

Mumbai City: Geographical Location of Research

Table 4.1: Statistics of Sampled Wards

Ward	*Population 2001*	*Per cent of Total City Population*	*Total Area (sq. km.)*	*Population Density (sq. km.)*	*Total No. of Election Wards*	*No of Reported Cable Connections*	*Linguistic Sample Units**
A	207,514	6.23	11.41	18,187	4	17749	G/M
C	190,672	5.73	1.78	1,07,119	4	23220	Gujarati
D	378,607	11.38	6.63	57,105	9	12132	G/M
F/N	526,839	15.84	12.98	40,588	13	12429	Marathi
G/N	590,609	17.75	9.07	65,227	10	18002	G/M

* G/M = Gujarati as well as Marathi

Each administrative ward consisted of a number of election wards. At random, one election ward was selected from each administrative ward. It was followed by the selection of the geographical location in each Ward. Selection of 50 Gujarati or 50 Marathi C&S TV households in each ward was not a difficult task from administrative Wards having average population density as high as 45,989 persons per sq. km.

Following factors were considered in finalizing geographical location:

- Respective language medium school in the vicinity.
- People especially women in culture-specific clothing or speaking in their respective languages and were moving on streets if responded positively became contact.
- Some local contacts who can help locate first Gujarati or Marathi household.
- Local shopkeepers, cable operators information about predominance of a particular linguistic group in the geographical location.
- Contact of a person provided by the previous viewer.

Table 4.2: Sampling Frame

Mother tongue	*Ward*	*Sex*		*Total*
		Female	*Male*	
Gujarati	A	25	25	50
	C	25	25	50
	D	25	25	50
	G/N	25	25	50
	Total	100	100	200
Marathi	A	25	25	50
	D	25	25	50
	F/N	25	25	50
	G/N	25	25	50
	Total	100	100	200

Only one person from a household was interviewed to avoid repetition about family-related data and to increase coverage of the research. In order to get diverse geo-graphical coverage not more than five households were interviewed from a given Housing Societies/*Chawl*/Flats/Slums. The present research covered 80 Gujarati households from Housing Societies/*Chawls*/Flats and 91 of those for Marathi households. Table 4.3 indicates Ward-wise and total coverage of the research across mother tongue.

Table 4.3: Geographical Distribution of the Sample

Mother tongue *Administrative Ward*	*Number of*			
	Gujarati Speaking		*Marathi Speaking*	
	Geographical Locations	*Housing societies/ Chawls/flats*	*Geog-raphical Locations*	*Housing societies/ Chawls/flats*
A	7	25	8	23
C	5	27	—	—
D	8	22	6	24
F/N	—	—	6	17
G/N	5	26	4	27
Total Administrative Wards		4		4
Total Election Wards covered		14		14
Total Geographical Locations		25		24
Total Housing societies/ *Chawl*/Flats/Slums		80		91

Family Group Interviews: At the end of the individual interview each viewer was asked to mention his/her willingness to be part of 'family group interviews' and telephone number, if they had telephone. Analysis showed that 115 (57.5 per cent) Gujarati and 99 (49.5 per cent) Marathi viewers expressed their willingness for family group interview.

Considering data collection experiences at the time of individual interviews, 'prior appointment' was considered essential for family group interviews. Since the proportion of households having telephone connections was high, telephonic appointments were fixed for family group interviews. Though many viewers had given their telephone numbers at the time of individual interviews, when actually contacted they expressed their inability to participate in family group interview. There were varied reasons like 'would not want to repeat interviews', 'spouse refusal', 'lack of time', and 'unsuitability of time for all the family members'. Among those who agreed for family group interview, further selection was made keeping family's socio-economic status, family composition and nature of television viewing reported by the individual viewer during the survey.

Table 4.4: Sample Distribution of Family Group Interviews

Ward / *Mother Tongue*	*A*	*C*	*D*	*F/N*	*G/N*	*Total Families*	*Number of Family Members*	
							Total	*Co-vered*
Gujarati	2	2	2	—	2	8	42	36
Marathi	2	—	2	2	2	8	41	38
Inter-community	—	—	—	1	—	1	3	3

Table 4.4 illustrates the coverage of family group samples across geographic wards, family size and number of family members present for the family group interviews. Table 4.4 indicates that of the total 86 members 77 of them participated for the family group interviews with the average family size of

five. Table 4.5 shows family structures of the interviewed family groups.

Table 4.5: Family Structures of the Interviewed Families

Gujaratis	*Marathis*
DAULATJADAs: Retired Father + mother, unmarried daughter, working son, his wife and grandchild (3 months).	**PATANKARs:** Grandparents, son-wife and granddaughter (7).
VYASs: Grandparents, two sons and their wives, granddaughter (18) of deceased daughter-in-law and grandson (4).	**TRIPADs:** Grandparents, 2 unmarried daughters, 2 sons and their wives, 3 school-going grandsons and a niece.
NISARs: Grandparents, son-wife and grandson (12), widowed daughter-in-law and two grand-sons (4½ and 8).	**VELUNKARs:** Grandparents, widowed daughter-in-law and grandson (21).
VASHIs: Widowed grand-mother, father + mother, son (13), daughter (16), relative's son (19).	**KADAMs:** Widowed grand-mother, father + mother, son (7), daughter (2 months),
KADAKIYAs: Father + mother, two unmarried working daughters.	**BALIGAs:** Father+ mother, two daughters (12 and 20).
PATELs: Father + mother, two unmarried working sons.	**MULAYs:** Father + mother, two unmarried working sons, (visiting) married daughter.
ASHARs: Father + mother and son (10)	**LAGADs:** Father + mother and daughter (10)
MODYs: Sister-in-law, mother and unmarried working daughter.	**AGARKARs:** Father + mother and daughter (10)
Intercommunity Household— **MEHTAs:** Father + mother and son (18)	

Age of the Individual

Nature of family relationships across households covered for the family group interviews varied as shown in Table 4.5.

VISUAL-2: Gujarati Households

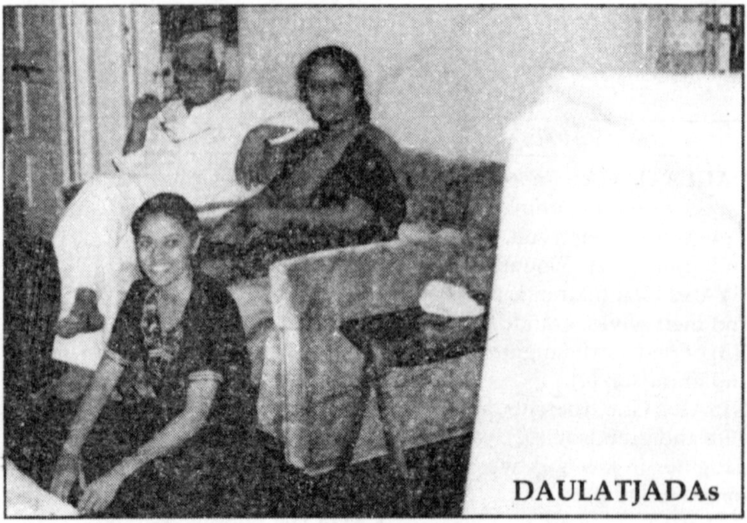

DAULATJADAs

VYASs

NISARs

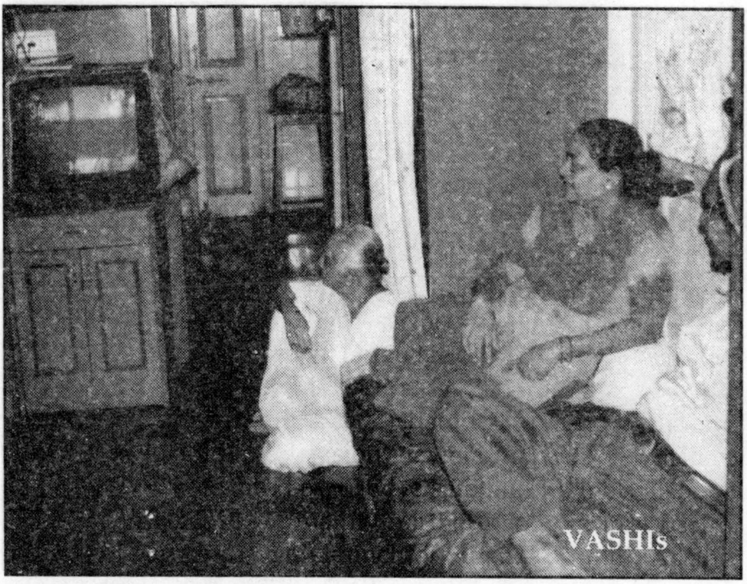
VASHIs

KADAKIYAs

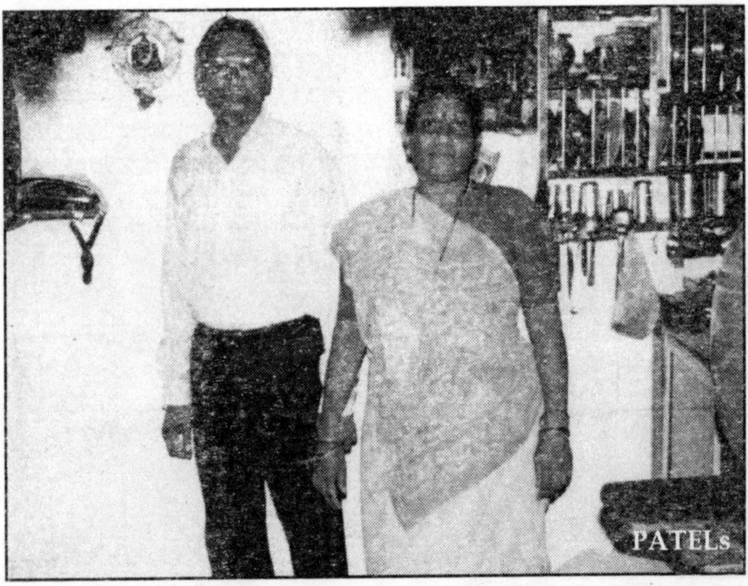

PATELs

ASHARs

MODYs

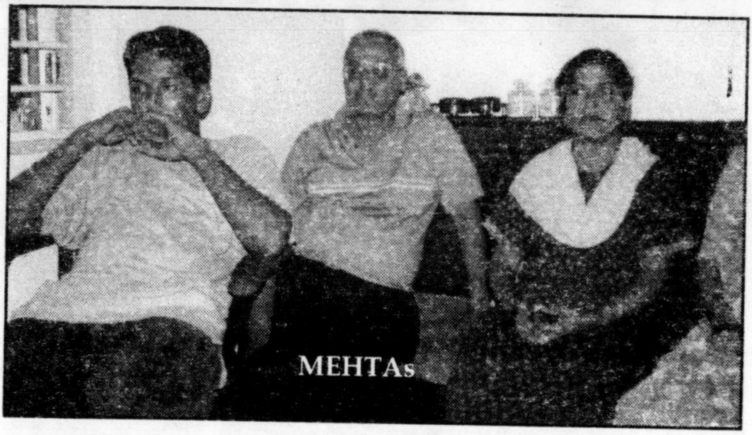
MEHTAs

VISUAL-3: Marathi Households

PATANKARs

TRIPADs

VELUNKARs

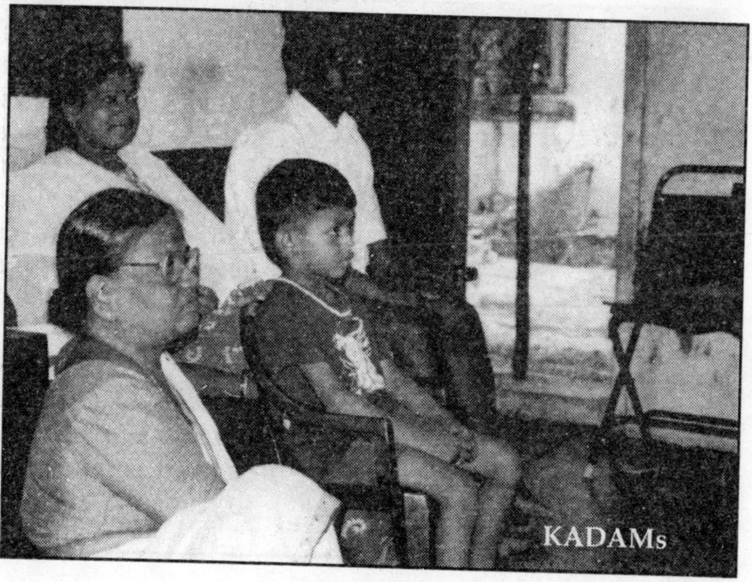
KADAMs

BALIGAs

MULAYs

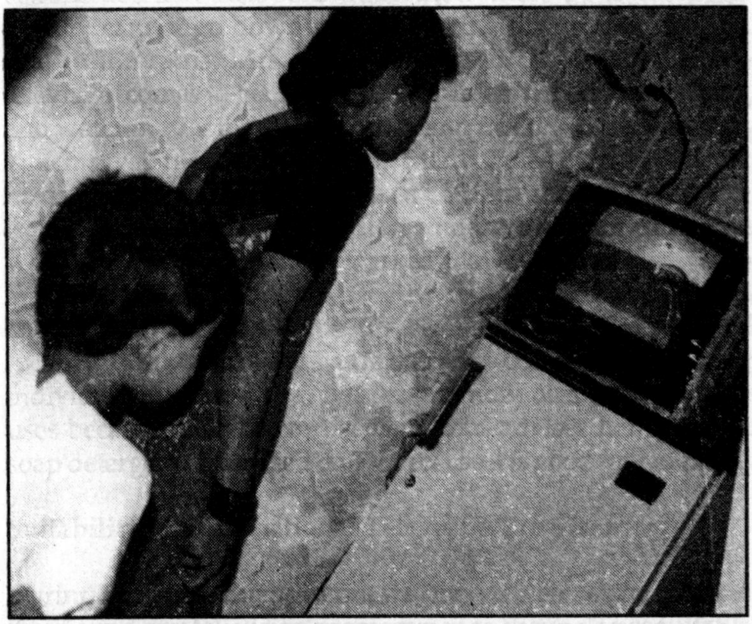

The household compositions were:

- Nuclear family with husband, wife with/without children (only sons/only daughters/single child/son or daughter or both)
- Joint family of grandparents (both or either of them) with young couple with/without children. Nuclear family having relatives' children or extended family of that type
- Joint family having brothers' families
- Grandparents with widowed daughter-in-law/s and grand-children
- Parents with adult unmarried daughters/sons
- Grandparents with unmarried daughter and son's family

Data Collection

Two separate instruments were used for data collection. These included 'individual interview schedule' having scales on individualism and consumerism and 'family group interview schedule'. Each has been separately discussed in detail.

Individual Interview Schedule: Review of the literature and experiences of prior research indicated that tools prepared in two different languages might not yield comparable analysis and consequently affect the findings of the research. Hence, it was decided to construct all the tools in English and later translate them in respective languages—Marathi and Gujarati.

Construction of Scales: The construction of scales were considered important since the research meant to examine influence of transnational television on values of individualism and consumerism. It was decided to develop separate scales on 'individualism' and 'consumerism' so as to assess these values separately. Based on the extensive review of related literature operational definitions of individualism and consumerism were worked out. Once the operational definitions were reached, two lists of the selected items were constructed. From the universe of a large number of items 45 items each for 'individualism' and 'consumerism' were selected for pilot testing.

Since the first draft scale items were in English the task was to translate them in Gujarati and Marathi. The exercise of validation of translation was carried out using 'back translation method' (Brisil, 1976: 154).

The draft of 45 items each on individualism and consumerism were sent to 40 experts from diverse disciplines of which 30 responded by rating items. The scales were also administered to 40 viewers (equal proportion of Gujarati and Marathi speakers) who were not the part of the final sample. Item analysis of the viewers and the experts' responses was carried out. After item analysis 20 items on individualism and 25 on consumerism were selected [Annexure A(i) and A(ii)].

Prior to finalization of data collection tool, a pilot survey was carried out. Forty viewers of C&S TV households (equal number of Gujarati and Marathi) spread across 20 geographical locations of Mumbai were purposively selected and interviewed. While many viewers preferred filling up the interview schedules, others preferred to give face-to-face interview. In the light of pilot testing, some questions were modified and few were added. The interview schedule after pilot testing was finalized in Gujarati and Marathi languages. In all, there were 43 questions and 46 sub-questions besides 45 scale items. The data collection instrument also included a covering letter stating the purpose of the research, instructions for filling up and an end-note with a request "if the researcher is permitted to come for a family group interview afterwards please provide telephone number, if any". Reason for keeping covering letter, based on pilot study experiences, was to allow viewers fill it up on their own, if they do not have tune for the interview.

Family Group Interview Schedule: A semi-structured interview schedule was constructed to document viewing behaviour of the family. It was used as an interview guide but each family group interview brought out specific point of interest for that household. It had four main areas and 21 sub-areas of inquiry related to television set, set usage, programme choices and areas of influence of television.

Data Collection Procedure: The data was collected during last week of December 2001 to second week of June 2002. About 70 per cent of the data collection had to be done through personal interviews while remaining preferred to fill up the interview schedule on their own. The individual interviews as well as household interviews were conducted mostly in respective mother-tongues and if needed in English depending upon their language preference. Total 44 per cent of the viewers (29 per cent Gujaratis and 15 per cent Marathis) answered the interview in English because they were not able to or wish to speak/write in their respective mother tongue. Most of them were in 20 to 30 years of age group. Individual interviews took on an average 45 minutes ranging from minimum of half an hour to maximum of an hour and a half.

The family group interviews were conducted in June 2002. The family group interviews lasted for about an hour and a half. Since it was found difficult to get all family members at one time in few households, few members could not participate in family group interviews. Except six Gujarati and three Marathi household family members in all total 77 individuals participated in family group interviews as shown in Table 4.4 and Table 4.5 earlier. Due to insufficient space in few cases, all family members could not sit together and participate in the family group interview. In some cases, the group interaction was intense and all family members conversed at the same time whereas in few cases they spoke 'one after another' or 'to each other' in the presence of the researcher.

Majority of the family group interviews were conducted at night after 8.00 p.m. since that was family get-together and television viewing time. Few interviews were conducted on holiday/Sunday afternoons depending upon convenience expressed by the family.

Data Collection Experiences

Present research covered 400 viewers of two linguistic communities having an average family size of 5.2 persons. Viewers had expressed inherent difficulty of 'time' considering the distances. Following are the highlights of the data collection experiences.

- On an average more than one visit was made to complete an interview. Majority of the viewers initially stated that they did not watch much of television as television was not important part of their life. However, observations indicated that many of the viewers were actually 'heavy' television viewers. Often the viewers were apologetic about their television viewing behaviour. Female viewers in spite of their other priorities were cooperative and spared sufficient time with the researcher. On the other hand male viewers were found to be difficult to interview due to their long working hours. So, many male viewers were interviewed at their work place or they filled up interview schedule in the presence of the researcher or at times completed schedule was collected by the researcher.

- Family group interviews were easier to conduct due to the established rapport with individual member of the household. Most of the families began the conversation by saying that they did not watch much of television. It allowed first hand observation about varied nature and type of 'household' as the 'family'. Family group interviews provided deeper insights into viewers' socio-cultural milieu and households' usage of medium of television.

Data Analysis

Data was analyzed using Statistical Package for Social Sciences (SPSS) version 10.1. Since most of the variables showed normal distribution, parametric tests and t-tests have been applied to assess test of significance especially for the values of individualism and consumerism and daily average time spent on television. Measures of association using Pearson's Correlation Coefficients at 95 per cent confidence interval have been used to explore variables influencing transnational television viewing and its association with various other independent and dependent variables. Annexure-B gives operational definitions of the concepts of the research and Annexure-D elaborates the indices used for better presentation and meaningful analysis of the findings.

5

Television and its Viewers

In India television viewing is still a 'family' activity, therefore, the profile of the viewers as individuals as well as in terms of household is of equal importance. Household is "a group of persons who commonly live together and would take their meals from a common kitchen unless the exigencies of work prevent them from doing so. The member of the household may be related by blood, un-related or having a mix of both", as defined for the present research. Viewer referred to "any individual of either sex residing in Cable and Satellite television household in the City of Mumbai between the age group of 20 to 50 years".

Household Profile

Household profile includes family size, family composition, education and occupation of the head of the household, ownership of household appliances and lifestyle products, membership of community organization and language spoken at home.

Family Size

The households covered for the research had an average family size of five members with higher variance (6.3) among Marathi households compared to Gujarati households (5.3). Total number of members in the 400 households covered was 2042. Almost seven out of ten households had three to six members in their family. About ten per cent of Gujarati households had seven-eight members while similar proportion of Marathi households had more than eight members. Interestingly Marathi as well as Gujarati households covered for the present research had positive sex ratio.

Family Composition

Household member data was collected in terms of three categories: children (below 12 years of age), adolescents (between 13 to 18 years) and adult members (above 18 years of age). Sex and number of family members in each category was asked to understand nature of family composition. Almost four out of ten households had 'only adult' members of which four out of ten were Gujarati households compared to three out of ten Marathi households. Since the viewers belonged to 20 to 50 years of age, more of the families were young families as little less than one-third had 'adults and children' members while about one-fifth had 'adults and adolescents'. Ten per cent of the households across mother-tongue had children as well as adolescents.

Nature of family composition in the household varied as emerged during the family group interviews. Broadly a large number of the households were nuclear followed by joint family composition. Few households reported to have relative's children staying with the family for studies or work purpose.

Education and Occupation of the Head of the Household

The viewers were asked about education and occupation of their head of the household. It was observed that viewers across mother-tongue most often named the eldest member of the household (old grandmother or retired father/mother) as the head of the household. This clearly indicates Indian family hierarchy of age and relationship. In some cases, sole-earning member of the house was considered 'head' of the household. Almost half of the head of the households were 'undergraduates or graduates' followed by about two-fifth educated 'up to SSC or below'. Little more than ten per cent households had postgraduate head of the households. There was not much of difference across mother-tongue except proportion of illiterate heads was double in Marathi households compared to Gujarati.

Occupational profile of head of the household reflected traditions and occupational profiles linked to linguistic backgrounds. One-fifth of the households reported 'old mothers or retired father/mother' as the head of the households remaining heads of the households were occupationally active. Close to half of the Gujarati head of the households were shopkeeper/trader or were in business compared to fifteen per cent of Marathi households. Marathi head of the households were working as clerks, officers, salesmen, supervisor or officer compared to much lesser proportion of Gujaratis in that category. About ten per cent of Gujarati household heads were 'self-employed' as against five per cent of Marathis.

Ownership of Home Appliances and Life-style Products

Question related to ownership of home appliances covered ten appliances: telephone, VCR/VCP/VCD, two-in-one, music system, washing machine, fridge, air conditioner, two-wheeler except cycle, car and computer. Telephone and refrigerator were most frequently owned as eight out of ten households owned them. Six out of ten households owned washing machine and five out of ten households owned music system. More than one-third (37.3) households owned computers while sixteen per cent had Internet connection.

Gujarati households owned many more appliances compared to Marathis indicating better socio-economic profile or adaptability of technology among Gujaratis. This was also evident from the fact that more than half of Gujarati households had computer as against one-fifth of Marathi households. About one-third of the Gujarati households owned vehicles like two-wheelers or car compared to only ten per cent of Marathi households. When asked to report 'any other' appliances most viewers reported grinder, iron or cycle while few high-income viewers reported ownership of Microwave oven or juicer. Ownership of all appliances except telephone, two-in-one, washing machine, and refrigerator was in the proportion of 2:1 in favour of Gujaratis.

Home appliance ownership data was consolidated to get socio-economic profiles. As evident from earlier discussion, Gujarati households owned higher number of appliances compared to Marathi households. Little over two-fifth of the Marathi households had 'three or less' appliances while similar proportion of Gujarati households had 'more than six' appliances.

Life-style Products

Apart from home appliances viewers were asked if any member of the household owned lifestyle products like laptop computer, credit card, mobile phone or club membership. Four out of ten households owned mobile phone and/or credit card with sharp differences across mother tongue. Laptop computer was owned by only three per cent of the households but many households reported ownership of personal desktop computer as discussed earlier. On the whole, 16 per cent of the households had membership of swimming pool or recreational club.

Similar to the home appliances, Gujarati households owned higher number of lifestyle products as compared to Marathi households. Analysis of number of life-style products suggested clear differences across mother tongue. Majority of the households owned one product or the other as only two-fifth did not own any products of which seven out of ten were Marathi households and two out of ten were Gujarati households. Proportion of Gujarati households owning more than two lifestyle products was 2:1 compared to Marathi.

Household Socio-Economic Status (SES)

Household SES was calculated by consolidating 18 items as discussed in the Annexure-C. Close to half of the Marathi households belonged to SES-C compared to only fifteen per cent Gujarati households. Five out of ten Gujarati households belonged to SES-A as against two out of ten Marathi households. Proportion of SES-B Gujarati and Marathi households was 4:3. SES comparative analysis is in conformity with the earlier discussion about higher degree of ownership of household appliances and lifestyle products among Gujaratis.

Membership to Community Organization

Viewers were asked about membership to community organization so as to assess their social affiliations. While more than half of Gujarati households were members of their respective community organizations irrespective of their socio-economic status, only one-fourth Marathi households reported such membership. Among Marathis, higher proportion of middle SES Marathis were members of their community organization compared to lower or upper SES. Historically Gujaratis having migrated from Gujarat and in the need to be 'together' and maintain 'community identity' reported such social affiliations compared to Marathis. Besides, "as the world becomes an electronic village, world's people cling even more strongly to their native cultural identity" (Doyle, 1992: 98). Gujaratis in Mumbai are away from their home state and, therefore, might be feeling the need to be having their community affiliations.

Language Spoken at Home

The viewers were asked to state the language which they spoke 'most frequently' at home. Nine out of ten households reported the language 'most frequently spoken at home' as their mother tongue. Only 7.5 per cent used English in combination with their mother tongue while 2.5 per cent communicated in multiple Indian languages like Konkani, Hindi or Kannada at home. Marathis were found to be having higher affinity towards mother tongue compared to Gujaratis as 86.5 per cent of Gujaratis conversed in Gujarati (including 4.5 per cent *Kutchi*) compared to 93.5 per cent of Marathis converse in Marathi at home. One-tenth (10.5 per cent) of Gujarati households conversed in English at home compared to 4.5 per cent of Marathi households. Gujarati viewers explained that they converse in English to improve the language of their English medium school-going children.

Analysis of Socio-Economic Status and language indicated that with the rise in SES (from SES-C to SES-A), use of mother tongue had reduced (95.5 per cent of SES-C viewers compared to 82.5 per cent of SES-A viewers conversed in their mother tongue at home). Viewers reporting multiple Indian languages usage to converse at home, mainly belonged SES-C households.

Profile of the Viewers

Profile of viewers refers to age, education, occupation, religion, marital status, years of residence in Mumbai, language efficiency and media usage with reference to 'print media' (newspapers, magazines, books), 'electronic media' (radio, television, VCP/VCR/VCD), 'convergence media' (Computer and Internet) and media usage outside the house (Films in cinema halls, plays, Internet access).

Age

The present research was carried out among the age group of 20 to 50 years as discussed earlier. The average age of the viewers was 32 years and the deviation of age in the total sample was nine years four months. Half of the viewers were in the age group of 20 to 30 years of age. There was not much of variation across mother tongue except for 20 to 30 years of age as 56 per cent of Marathi viewers and 45.5 per cent of Gujarati viewers belonged to that age group. Skewness of age distribution was 0.38 indicating that the data was normally distributed.

Religion

More than three-fourth of the viewers were Hindus across mother tongue and their proportion was higher among Marathi (80 per cent). One-fifth of Marathi viewers were Baudh. One-fourth of Gujaratis were mainly Jains, few Parasis or Muslims. Marathis were either Hindu or Baudh as against multiple religious backgrounds of Gujaratis like Hindu, Jain, Parasi or Islam.

Marital Status

Six out of ten viewers were married and as obvious almost nine out of ten of the unmarried viewers were in the age group of 20 to 30 years. Analysis indicates lesser proportion of married Marathis in younger age groups compared to Gujaratis and, therefore, more Gujaratis were married compared to Marathis.

Education

Educational status of the viewers across sex and mother tongue reveals that half the viewers were under-graduates and one fifth were post-graduates. There was no sex difference in educational status. However, there were striking differences amongst males across mother tongue. More than two-fifth of the Marathi males had studied up to 'SSC/HSC or below' compared to about one-fifth of the Gujarati males. Proportion of Gujarati post-graduates was double than Marathis. Only three Marathi females reported to be illiterate while not a single Gujarati was illiterate of the total sample of 200 thereby indicating almost 100 per cent literacy among the C&S TV viewers.

When examined in relation to the educational status of the head of the household education of the viewers clearly suggested that the present generation of Gujaratis had attained better educational status compared to their Marathi counterparts yet unlike their respective head of the households they had higher educational levels.

Occupation

Majority of the viewers were in job/service yet having differences across sex and mother tongue. Almost half of the females were housewives and one-third were in service compared to six out of ten males who were in service/ professional jobs. Equal proportions of viewers across sex were students considering the age profile of the viewers. Occupationally one-fourth of the males were into 'business or independent work' compared to only ten per cent females.

'Business/independent work' category included viewers pursuing paid housework or home-based business like running childcare centre, tailoring work, taking tuition of dance-music or school children, LIC agency work, web designing work or even helping spouse's office work from home. Few viewers also reported to be involved in multiple occupations like business as well as job, or consultancy and job, or tuition and job.

Close to half of the Gujarati males were in business/ independent work compared to only fourteen per cent Marathis. Three out of four of the Marathi males were in service/ profession compared to two-fifth of Gujarati males. The occupational pattern of Gujaratis and Marathis indicated prevalence of family business among Gujaratis and service-orientedness among Marathis. More than half of the Gujarati females were housewives as against 38 per cent of Marathi females. Proportion of 'professionals' among Gujaratis was double than that of Marathis mainly due to higher educational attainments.

Student population was only in 20 to 30 years of age group which included total of one-third Gujaratis and one-fifth Marathis. More than half of the Marathis were in service in 20 to 30 years of age group compared to about one-third of Gujaratis. In 30 to 40 years of age group proportion of Marathis in service was almost double than Gujaratis. Data indicated that Marathis start working at an early age and marry at a later age compared to Gujaratis, which could be function of better economic support provided by Gujarati parents. It may also reflect need for the Marathi youth to start working at a young age to be economically independent early in life since the occupational dependence is mainly on 'service'. The sample distribution also confirmed the popular belief that the Gujaratis are business people and more of Marathi females are working women.

Residence in Mumbai

The viewers were asked, for how long they have been residing in Mumbai so as to ascertain whether she/he was resident or migrant. The data of years of residence in Mumbai was cross-tabulated with the age of the viewers to ascertain the migration pattern. Almost nine out of ten viewers were residing in Mumbai 'for generations' with little difference across mother tongue. Many of them reported that they had been staying in the same locality or even in the same house for all those years. One Marathi household (Tripads) covered for the family group interview was fifth generation and another Gujarati household (Modys) was third generation residing in Mumbai. About ten per cent of 'first generation' residence of Mumbai reporting 'last few years' of stay in Mumbai stated period of five to fifteen years included five per cent of Marathis who migrated from other parts of Maharashtra mainly because of their jobs. Hardly any Gujarati covered in the research had come to Mumbai in last few years. Also "industrial development in Gujarat has blocked the flow of labour as well as the trading class to Bombay [Mumbai]" (Mallison, 1995:95).

Interestingly, 70.4 per cent of the Marathis in the age group of 20 to 30 years were first generation Mumbai residents compared to 47.6 per cent of Gujaratis in that age group. But in 40 to 50 years age group more than one-fourth of the Gujaratis were first generation residences compared to one-tenth of Marathis, suggesting differential pattern across Gujarati and Marathis. It must be mentioned that migrants coming to Mumbai initially settled in Mumbai City, later on shifted to sub-urban areas. Therefore, such large proportion of viewers who claimed to be residing in the city for generations and could have settled there since their entry into the city.

Language Profile

Viewers' spoken language was considered as an important dimension of the present research. Language included the medium of instruction at school, familiarity with English and language efficiency profile of the viewers.

Each of the viewers was asked to name 'languages' she/he can 'read, write, speak', so as to ascertain language efficiency profile of viewers across mother tongue. Language efficiency index was constructed as discussed in Annexure-C. The viewers indicated varied degrees of competence in eight Indian (Gujarati, Marathi, Hindi, Bengali, Tamil and Kannada, Konkani, one dialect *Kutchi*) and four foreign (English, French, Russian, German) languages. The analysis indicated that except for 0.05 per cent all viewers were multilingual having different degrees of competence in more than one language.

Medium of Instruction at School

Majority (68 per cent) of the viewers' medium of instruction at school was their mother tongue. About one-third (30.5 per cent) of the viewers studied in English medium. Two per cent viewers included three illiterate Marathi females as well as those having studied in Kannada and Hindi medium schools.

More than three-fourth Marathi viewers compared to four-fifth of the Gujaratis studied in their respective mother tongue at school level. Proportions of Gujaratis having studied in English medium schools was double than their Marathi counterparts.

Analysis of age and medium of instruction at school indicated that Marathi viewers had studied in their mother tongue irrespective of their age, while higher proportion of Gujarati viewers opted for English medium schools over the years. More than half (57.1 per cent) of the Gujarati viewers in 20 to 30 years of age had studied in English medium compared to only one-fifth in other age groups.

Mallison comments that "after 1960, when a separate Gujarat State was created, Gujarati became a minority language in Bombay [Mumbai]. This may be one of the reasons why Gujaratis in Bombay began to switch over to English-medium schools, the way other linguistic minorities such as Tamilians and Punjabis had been doing" (Mallison, 1995: 95).

About one-fourth (28.9 per cent) of Marathi viewers in 40 to 50 years of age group studied in English medium as against one-fifth (20.8 per cent) of Gujaratis, suggesting higher proportion of older Gujaratis compared to Marathis who had studied in English medium. Contrary to that large number of children irrespective of economic status and mother tongue of the household reported to be going to English medium schools, now.

Analysis of SES and medium of instruction at school in the light of mother tongue indicated inverse relationship between SES and mother tongue as medium of instruction at school. In other words, as SES increases, the education in mother tongue school decreases. Majority (89.6 per cent) of the SES-C viewers studied in their respective mother tongue compared to 69.6 per cent SES-B and 44.3 SES-A viewers.

Language Efficiency

Language efficiency index was constructed as elaborated in the Annexure-C. Across mother tongue more than half (55 per cent) of the viewers had 'medium' language efficiency. Only about two out of ten (17.3 per cent) viewers were found to be having 'low' language efficiency while 27.5 per cent had 'high' language efficiency, might be due to multilingual environment of Mumbai City.

Two-fifth (42.5 per cent) of Gujaratis had 'high' language efficiency compared to only 12.5 per cent Marathis and over one-fourth (28 per cent) of Marathis had 'low' language efficiency compared to only 6.5 per cent Gujaratis. The analysis confirmed that higher educational attainment had higher language efficiency and males had better language efficiency compared to females.

Familiarity with English

Eight out of ten viewers (81.8 per cent) reported familiarity with English at either or all the levels of reading, writing, speaking of which proportion of Gujaratis (90 per cent) was higher than Marathis (74 per cent). When familiarity with English was compared with the language efficiency scores, it was found that viewers having familiarity with English had better language efficiency scores.

Media Habits of the Viewers

Media usage of the viewers was analysed in terms of the frequency of media usage, language of media usage, and names and nature of media used inside and outside the household by the viewers. The analysis covered usage of print media, electronic media, convergence media and outdoor media like films and theatre.

At least six out of ten viewers used print media in form of newspapers and magazines and outdoor media in form of films in cinema halls and plays. Apart from television, the most used in-house media was newspaper (94.5) across mother tongue. More than half (54.4 per cent) of the Marathis read books while seven out of ten (65 per cent) Gujaratis read magazines. Higher proportion of Gujaratis used electronic and convergence media and outdoor media in form of films in cinema halls compared to Marathis. This could be due to the higher SES and ownership of appliances among Gujaratis. Newspapers were read 'almost everyday'; magazines and books were read 'once a week'. Proportion of non-users of newspapers, magazines and books (print media) were highest in 20 to 30 years of age group compared to other age groups. Age group of 20 to 30 years used the print media the least including books and depended more upon convergence media of computer and Internet.

All the viewers had access to television considering the objectives and sampling of the present research. In terms of frequency of usage, radio and television were listened to/viewed 'almost everyday' but VCR/VCP/VCD usage was limited to 'once a month' irrespective the age of the viewers. Age group of 30 to 40 years used electronic media marginally less than the other age groups while the usage was marginally higher among 40 to 50 years of age group.

Convergence media like computer and Internet were mostly used by young age group of 20 to 30 years. About one-third (31.6 per cent) of viewers in 40 to 50 years age group used computer compared to 48.3 per cent of 20 to 30 years age group. Similarly, Internet was used by 43.3 per cent of 20 to 30 years old compared to 22.4 per cent of 40 to 50 years age group.

There seems to be inverse relationship between age and convergence media usage. Analysis with sex also revealed that males used more of convergence media compared to females.

Media Usage Outside the House

Apart from films and plays, convergence media (Computer or Internet used at work, research place or cyber café) was analysed as outside house media usage. Film viewing in cinema halls was found to be very popular mainly among Gujaratis followed by plays and convergence media usage. More than one-fourth of the viewers had watched film in cinema halls in last fortnight. Viewers in 30 to 40 years age group visited cinema halls and watched plays more frequently compared to other age groups. But if data are pooled for the whole month, majority (63.1 per cent) of the 20 to 30 years age group had watched film in last one month, indicating high frequency of usage of film medium by them compared to other age groups.

As discussed earlier, 37.3 per cent of the households owned computers but the proportion of those using it was 42.9 per cent. About 16 per cent of the total households had Internet connection while one-third (33.5 per cent) viewers reported Internet access at home or outside their house.

Language of Media Usage

Analysis suggested that the viewers used media in various languages. The media usage was invariably multilingual as most of the viewers were multilingual as discussed earlier. Tables 1 and 2 (Annexure-E) show language preferences of viewers across print and audio-visual media.

Table 1 shows that more than half (51 per cent) of Marathis read newspaper in their mother tongue compared to about one-third (34.5 per cent) of Gujaratis. Less than one-third (29 per cent) of Gujaratis read English language newspapers and about equal proportion (32.5 per cent) of them read them in combination with Gujarati newspaper. About one-fourth of Gujaratis read magazines in English or Gujarati while 13.5 per cent read in combination of both the languages compared to 29 per cent of Marathis who read Marathi magazines. The other Indian language of print media usage was Hindi or occasionally Marathi. Few Gujaratis who were born and brought up in Mumbai and even studied Marathi as second language, having familiarity with the language reported usage of Marathi language media.

Table 2 (Annexure-E) shows that Hindi emerged as an important language of media consumption as far as films and television was concerned unlike English in print media. 'Only English' language films and television was viewed by less than ten per cent of the viewers. But the proportion of viewers watching Indian as well as English language content were close to one-third for films and about one-sixth for television across mother tongue. Tables 1 and 2 (Annexure-E) clearly show that English newspapers and magazines, Hindi films and television and Marathi books and plays were accessed by higher proportion of viewers.

Analysis of language of media usage suggested that viewers were exposed to English language content in various media apart from television. Table 3 (Annexure-E) clearly shows that Indian language media consumption dominated the audio-visual media especially television but usage of English print media alone and in combination with Indian language media was also very high.

Nature of Individual Media Usage

This section examines types and brand name of the individual medium used by the viewers and purpose of usage, so as to understand variations in media usage.

Newspaper Reading

Viewers named 31 Newspapers including number of tabloids [Annexure D(i): p. 203] being read by them singularly or in combination with other newspapers. *The Times of India* was the most read newspaper by four out of ten (37.8 per cent) viewers of which one-fourth of the total viewers read only *'The Times of India'*. English newspapers were most often read in combination with Indian language newspapers as discussed earlier (*See* Table 1 Annexure-E). Gujaratis read more of *Mumbai Samachar* or *Gujarat Samachar* and Marathis read *Loksata* or *Maharashtra Times*. Language readership of newspaper *vis-à-vis* English newspaper was approximately 1:1 among Gujaratis and 1.5:1 among Marathis. Tabloids like *Mid Day* and *Samana* were found to be popular across mother tongue. Five viewers across mother tongue also reported to be reading Hindi newspaper *Nav Bharat Times*.

Magazine Readership

Viewers named 85 weekly or fortnightly magazines being read by them [Annexure D(ii): pp. 204-205]. In terms of readership 'general interest' magazines were most popular and were read by 64.1 per cent followed by women's magazine (27.3 per cent). In terms of number of magazines named by viewers 'special interest' magazine numbers were highest (29), which included variety of subjects like computer, photography, health, travel, textiles, sports, children's or youth focused magazines.

The general interest magazines mainly dealt with news and current affairs, which were read by slightly higher proportion of Gujaratis who also read religious, literary and film magazines. Women's magazines were read in equal proportions across mother tongue. Further analysis indicated that about one-third (32.7 per cent) of the magazine readers read multiple types of magazines.

Books Readership

Fiction more than non-fiction books were read by viewers. The proportion of Gujaratis was higher among fiction books readers while Marathis read non-fiction books. Unlike religious magazines that were read more by Gujaratis, equal proportions of viewers across mother tongue reported reading religious books.

Radio Listening

Half of the viewers did not listen to radio at all. Among radio listeners more than half listened mainly to music or news. Marathi viewers reported to be listening to radio stations like Mumbai A, Mumbai B and Mumbai-Pune while Gujaratis listened to Vividhbharti. Others included Ahmedabad, Delhi, BBC, and Ratnagiri Akashwani Kendra.

Usage of VCR/VCP/VCD

Most of the viewers owning VCP/VCR/VCD commented that they did not know if the equipment was working. Many viewers expressed that due to 24-hour television channels the VCP/VCR/VCD technology had lost its relevance. Besides VCD get used for at least new film but VCR/VCP are of no use as even the cassettes are not available in the market as stated by the viewers. In spite of the fact that 37.5 per cent of the total viewers owned VCR/VCP/VCD, only 24.5 per cent of them reported to be using it and that too for not more than "once a month". Marathi viewers were found to be using VCP/VCR/VCD for variety of purposes. Nine out of ten VCP/VCR/VCD viewers used to watch films, in which proportion of Gujaratis was higher than Marathis. Film viewing was followed by viewing of personal tapes (recordings of weddings, birthday party) besides other purposes like religious, science, Marathi *natak* (drama) viewing.

Convergence Media Usage

Internet was most frequently used for mail or chat and those owning computer and Internet at home used it for variety of purposes. Table 4 (Annexure-E) clearly showed that ownership of computer and Internet at home was directly correlated with its usage.

Analysis of media usage by sex revealed that in general, males used more of newspaper and convergence media while females used books and magazines. But in the context of mother tongue more of Marathi males read books and more of Gujarati males used Internet. More of Marathis youngsters (20 to 30 years) used print media compared to their Gujarati counterparts who used more of convergence media (Tables not included).

Time Spent on Media

Table 5.1 shows that an average viewer spent higher amount of time with electronic media than print media except for reading books (55 minutes). Many viewers reported that they switched on television as soon as they get up or listen to radio even while travelling. The standard deviation in media usage time was much lower for newspapers (20.92 minutes) and much higher for television (105.6 minutes). The average amount of time spent on media was four hours and 24 minutes per day. Marathis spent higher amount of time with media compared to Gujaratis.

Table 5.1: Descriptive Statistics of Time (in Minutes) Spent on Media

Media	*Users N=400*	*Minimum (minutes)*	*Maximum (minutes)*	*Standard Deviation*	*Mean time (Approx. Minutes)*
Newspapers	377	0	180	20.92	30
Magazines	245	0	240	29.22	37
Book	191	0	300	39.39	55
Radio	199	0	540	57.59	63
Television	400	30	660	105.6	210

The mean time spent on media per day by Marathis was 4.7 hours compared to 4.1 hours spent by Gujaratis as shown in Table 5.2 with higher amount of deviation among Marathis. Close to half (46.5 per cent) of Marathis spend more than 4.5 hours with media compared to about one-third (31 per cent) Gujaratis. Similarly 41.5 per cent of Gujaratis were using media for less than 3.5 hours compared to 33.5 per cent Marathis. The difference in amount of time spent on media was higher among males across mother tongue. Gujaratis males (44 per cent) were spending less than 3.5 hours while half of the Marathis males were spending more than 4.5 hours. Comparatively Gujarati females and Marathi males spend higher amount of time with media, which also might have to do with occupation of viewers.

Table 5.2: Percentage Distribution of Time Spent on Media, Sex and Mother tongue

Amount of time spent per day	*Gujarati*			*Marathi*			*Total*
	Female	*Male*	*Total*	*Female*	Male	Total	
<3.5 hours	39.0	44.0	41.5	36.0	31.0	33.5	37.5
3.5 to 4.5 hours	29.0	26.0	27.5	21.0	19.0	20.0	23.8
>4.5 hours	32.0	30.0	31.0	43.0	50.0	46.5	38.8
Total	100.0	100.0	100.0	100.0	100.0	100.0	100.0
Mean		4.1			4.7		4.4
Std. Deviation		1.7			2.3		2.0

In Brief

Gujarati and Marathi speaking households indicated similarities in terms of positive sex ratio, education of head of the household and family size. The average size of the household was five. The family composition of Gujaratis as well as Marathi households was mainly either 'only adult' or 'adults and children'.

Among four out of ten head (most often the eldest member of the household) of the households across mother tongue had the education up to SSC/HSC or below. But the education level in general seems to have improved in the present generation especially among Gujarati viewers. Gujaratis were mainly in the business and trading while Marathis were most often in service.

Eight out of ten viewers had been residing in Mumbai for generations and the migration pattern clearly suggested that though little more than half of the Marathi viewers were in Mumbai for generations, none of the Gujarati viewers had come to Mumbai recently.

Gujarati viewers belonged to diverse religious backgrounds like Jain, Parasi or Muslim while Marathi viewers were either Hindu or Baudh. Gujarati households were found to be having higher number of home appliances and life style products compared to Marathi households. Half of the Marathi households were SES-C whereas similar proportion of Gujaratis households was SES-A, indicating better SES of Gujaratis. Higher proportion of Gujarati households irrespective of their SES were members of their respective community organizations as compared to Marathi households. Every six out of ten viewers were married. Marathis were found to be having higher age for marriage and lower age for work compared to Gujaratis.

Language spoken 'most frequently' at home was invariably mother tongue. However, Gujarati households were found to be using English more often than Marathi households. SES was found to be associated with English medium school education as well as English as language of conversation at home. Marathis viewers studied in their mother tongue medium schools irrespective of their age but proportion of Gujaratis in English medium schools had gone up over the years. Proportionately more of Gujaratis were found to be having familiarity with English and better language efficiency compared to Marathi viewers.

While viewers across mother tongue read newspapers, higher proportion of Marathis accessed print media in forms of books while Gujaratis read magazines. Gujaratis used electronic media, convergence media and film in cinema halls more than Marathis who preferred plays.

Language of media usage suggested that media consumption was multilingual. Most frequently print media (especially for newspaper and magazine) was used in English, audio-visual media (films and television) were accessed in Hindi. Viewers named 30 newspapers and 85 magazines being read by them. Usage of electronic media in form of radio and VCR/VCP/VCD was mainly for music and films. Convergence media especially Internet was accessed outside their house by one-third of the viewers for variety of purposes.

Analysis of media usage across age group showed that 20 to 30 years olds used print media the least and convergence media the most compared to other age groups. Age group of 30 to 40 years depended more upon media outside house like films and plays while usage of electronic media was marginally higher among 40 to 50 years age groups. Frequency of media usage showed that along with electronic media (except VCR/VCP/VCD) newspapers were read daily but magazines and books were read 'once a week'. VCR/VCP/VCD was used once a month.

Time spent on media suggested higher amount of time spent on electronic media and heavy dependence on television. The average amount of time spent on media was four hours and 24 minutes per day of which Marathis spent higher amount of time (4.7 hours) compared to Gujaratis (4.1 hours) per day. Comparatively Gujarati females and Marathi males spend higher amount of time with media.

6

Television in the Household

James Lull way back in 1988 remarked with reference to family viewing (of television) within the household, a complex mixture of people, social roles, power relations, routine activities, processes of interpersonal communication, ecological factors that characterize the home environment, the technological devices and appliances that exist there.

Present research examined nature of television (TV) ownership in the household covering issues related to type, size, number, channel capacity of TV set, years of cable connection, monthly expenditure for cable connection, reason for taking cable and if cable was disconnected at any time of the year. In the second part of the chapter, family TV viewing behaviour has been analysed covering television-viewing time of the family, decision-making about programmes, and channel surfing behaviour. Analysis is based on the findings of the individual survey as well as family group interviews.

Television in the Household

It was observed during data collection that TV set in the household was usually placed in the drawing room. For the houses having more than one room and households having multiple television set, the additional (most often smaller size TV set) was placed in the bedroom. Interestingly in all one room household TV set was placed at higher vintage point compared to multi-room apartments. It also seems to have association with economic background of the households. Lower SES households tend to put television sets higher for varied reasons like viewability, not to get it damaged due to space constraints or in a way display of 'importance' given to the medium in a day-to-day life of the household.

Table 6.1 shows the profile of the households covered during the family group interviews. As discussed in the Methods and Techniques Chapter 4, Family group interviews covered households having diverse SES, family size, television ownership purposively as evident from Table 6.1. Two households each across mother tongue having multiple television sets were covered through family group interviews. As Table 6.1 shows not a single Marathi household conversed in English compared to one Gujarati household (*Ashars*). Interestingly inter community household of Mehtas with Marathi female and Gujarati male having son of 18 years old conversed in Gujarati at home. The household in individual survey was considered as Gujarati household but during family group interview finalisation process was discovered as inter-community marriage.

Among Gujarati households Patels and Vyass were single-room houses while among Marathi households Tripads, Mulays and Kadams were single-room houses. The average family size of household covered for the family group interviews was also five. The average years of cable connection was higher among Gujarati households compared to Marathi households. Gujarati households had cable connection for more than five years as compared to four and half years among Marathi households. Table 6.1 shows that most of the households had acquired newer TV sets or had got channel converter in their TV to access multiple channels. Two households each (Daultjada and Nisar, Agarkar and Velunkar) had multiple TV sets.

Table 6.1: Profiles of households covered for family group interviews

Sl. No.	Geographical Ward	Family Name	Socio-Economic Status	Language at Home*	Family Size	Which no. TV set **	No. of Years of Cable	Head of Family's Education^
Gujarati Households								
1.	C	Vyas	Low	G	8	2nd	2	HSC
2.	C	Mody	Low	G	3	3rd	8	SSC
3.	D	# Daultjada	Upper	G	6	2nd +3rd	6	Graduate
4.	D	Kadakiya	Upper	G	4	2nd	6	Graduate
5.	G	# Nisar	Medium	Ku	8	1st Co+ 2nd	2	SSC
6.	G	Vashi	Upper	G	6	3rd	6	Graduate
7.	A	Patel	Medium	G	4	2nd	4	HSC
8.	A	Ashar	Upper	E&G	3	4th	8	P.P.G.
Marathi Households								
9.	A	Kadam	Low	M	5	2nd	4	HSC
10.	A	# Agarkar	Upper	M	3	1st +2nd	8	P.P.G.
11.	D	Mulay	Upper	M	5	1st Co	0.5	Graduate
12.	D	Tripad	Low	M	12	2nd	4	SSC
13.	F	Lagad	Low	M	3	1st	2	HSC
14.	F	# Velunkar	Upper	M	4	2nd Co+ 3rd	8	P.P.G.
15.	G	Patankar	Medium	M	5	3rd	4	Graduate
16.	G	Baliga	Upper	M & Ko	4	3rd	6	P.P.G.
Marathi Wife and Gujarati Husband								
17.	G	Mehta	Upper	G	3	4th	8	Graduate

\# Ownership of multiple television sets
^ P.P.G.= Professional Post Graduate
* Language spoken at home: M = Marathi, G = Gujarati, E = English, Ku = *Kutchi*, Ko = Konkani
** Number of TV Sets: Co + TV Set with Channel Converter

TV Set Ownership

Individual Survey revealed that nine out of ten households had colour TV set with remote control facility and eight out of ten households had only one TV set. Most of Gujarati households owned colour TV sets with remote-control facility compared to two of the ten Marathi households having Black and White TV sets and similar proportion without remote control facility. There were few Black and White TV sets with remote control. Some of the viewers reported that their original colour TV set did not have remote control facility having limited channel capacity. The set was modified by adding converter for allowing access to multiple channels. About one-fourth of Gujarati households had more than one TV set. Ratio of multiple TV set ownership among Gujarati and Marathi households was 2:1 (*See* Graph 6.1). Many households had opted for either bigger or smaller size sets, mostly to avoid tensions in viewing choices.

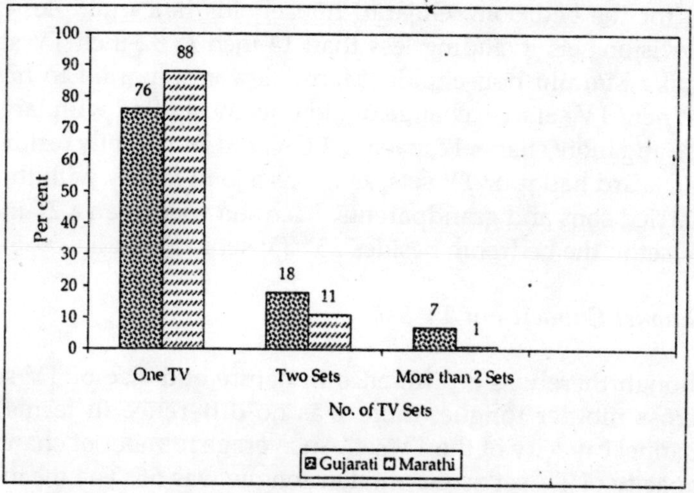

Graph 6.1: Ownership of Television Sets Across Mother tongue

Family group interviews revealed that most often second set was bought mainly to avoid conflicting viewing choices. The elders in Nisars household preferred Gujarati programmes while the others wanted to watch Hindi programmes. The new set, which was yet to be unpacked at the time of interview by the researcher would be kept in the youngest son's bedroom. Unlike that in Velunkars household second TV set was bought for the son who used to return home late and disturb other members of the house to watch television. After the death of the son, the TV set in the bedroom is used by the grandson.

Size of the TV Set

Individual survey revealed that seven out of ten households had 21-inch TV sets. Close to half (49.3 per cent) of the households owned 14-inch TV sets. The reason for large proportion of 14-inch TV set might be because of household space as most of the households were of smaller size and at times one-room accommodation. Close to one-fifth (18.5 per cent) of the total households had multiple television sets and their sizes varied. Most often in case of multiple sets ownership was of 21-inch set in the drawing room and 14-inch set for the bedroom. Gujarati households had multiple size television sets including less than 14-inch to 32-inch TV sets unlike Marathi households. Many viewers reported to have got new TV set in exchange of old one, most often with larger size and more channel capacity. A Gujarati joint family residing in C ward had four TV sets, as it was a joint family with three married sons and grandparents. Each son was given a 21-inch TV set in the bedroom besides 25" TV set in the hall.

Channel Capacity of TV Set

Though there was a difference in nature and size of TV sets across mother tongue, there was no difference in terms of channel capacity of the TV set. An average number of channel capacity of TV set across mother tongue was 56. The mean of channel capacity of television set in Gujarati households was 55.56 and SD was 39.13 whereas in Marathi households it was 56.29 with much higher SD of 43.59. The total SD of channel capacity was 41.19. The range of TV set channel capacity was from 6 to 250 channels. About one-fifth (17.5 per cent) of the viewers irrespective of their mother tongue had 'no idea' about channel capacity of their television sets.

Family group interviews gave insights into the processes related to first television set in the household. During the discussion about existing number of television set of the family it was found that most of the families reported entry of television in their households in 1980s or earlier with reference of some personal or national events like beginning of satellite broadcasting through INSAT, colour transmission, live telecast of Asiad Games (all in 1982) or Olympics or beginning of Bombay Doordarshan Kendra. Exactly half of the households covered through family group interviews had either parents or children who had viewed television at neighbour's place. Many families across communities reported their head of the household's dislike for the medium of television and as in case of Baligas television was bought after the couple got separated from the parents. Present generation of head of the households seemed to have accepted medium of television as many of them spend lot of time with it themselves.

All the households of family group interviews had gone in for modifications or change of television set after beginning of private satellite channels to access more channels. Families who did not wish to spend on new TV set added converter while others bought colour/with remote/bigger size or higher channel capacity TV sets. But like Vyass commented they wanted to take bigger size television set but there was no 'space' in the house. They stay in one-room household. Therefore, they bought 21-inch colour TV in their family home back in the hometown, which gets used in the vacations, mainly to fulfill their desire for larger size TV set. Their present TV set size was 14 inch.

Years of Cable Connection

The issue of years of cable connection was important because it made access to transnational television possible. During pilot testing when viewers were asked about the year of opting for the cable and satellite television, they found it difficult to specify the year instead they could respond in terms of time intervals.

Therefore, the question was asked using time intervals in the interview schedule. Besides unlike earlier cable connection, which allowed access only to local cable content, today's cable connection refers to access to Cable and Satellite television channels. Therefore, it has been referred as cable connection in the forthcoming discussion.

Most often viewers remarked having taken cable connection when Zee TV started that is in 1992 indicating ten years of cable connection at the time of individual's interview. The households having cable connection for longer years reported to have taken cable to watch films shown by cable operators. The recent adopters to cable connection were as recent as three months.

Four out of ten households opted for cable connection 'more than eight years back' while 9.5 per cent took cable connection 'in last two years'. About ten per cent of the households opted for cable connection each year in last two to six years. Six out of ten Gujarati households opted for cable connection more than eight years ago compared to 23 per cent of Marathi households. About one-fifth of Marathi households had taken cable connection in 'last two years'.

SES had association with adoption of cable connection. Years of cable connection and SES confirmed the findings of other studies (Kishore, 1994; Rahim, 1994; Gupta, 1998) that the early adopters of cable connection were upper SES households. More than half (52.2 per cent) of the households having cable connection for more than six years were SES-A compared to 69 per cent of households having cable for last two years were from SES-C. There was an inverse relationship between SES and years of cable connection.

Analysis of years of cable connection suggested that Marathis had opted for cable connection recently compared to Gujarati households. Independent sample t-test was employed to examine the statistical difference in years of cable connection across mother tongue. The years of cable connection among Gujarati and Marathi households differed significantly. But if one examines Table 6.1, the association of years of cable connection differed as among Marathis, Mulays though belonging to SES-A had opted for cable in last six months mainly for the daughter who had come for delivery. While among Gujaratis Modys from SES-A had cable since eight years.

Monthly Expenditure on Cable

The households spent monthly average of Rs. 125 against "national monthly average of Rs. 157" (Varghese, 2003). The cable monthly payments ranged from Rs. 75 to Rs. 500. The overall standard deviation of monthly payment was Rs. 101.4 with higher standard deviation (Rs.115.8) among Gujarati households compared to Marathi households (Rs. 87.8).

The ratio of viewers not knowing the amount paid for the cable connection was 2:1 between Gujaratis and Marathis. Some of the viewers reported that they were not knowing the amount paid by the household because either their employers or Housing society paid for the connection directly or it was paid yearly or quarterly and, therefore, they did not know the specific amount spent for cable connection every month.

Cable connection subscription across geographical ward varied depending upon the nature of housing or residential locality. Marathis residing in the slums in F ward paid Rs. 100 or less for cable connection while Gujaratis of A ward paid Rs. 250 or more for the same.

During the family group interviews, members of three households (Patels, Vyass, Mulays) expressed their concern for monthly expenditure on cable connection. Mulays (had taken cable connection in the last six months) reported that if present subscription would be raised they would disconnect the cable connection. Patels felt that they pay for television channels they do not watch. Contrary to that two Marathi households (Velunkars and Tripads) felt that cable was good entertainment at cheaper rate.

Reasons for Taking Cable Connection

Viewers were asked to give one reason as to why according to them cable connection was taken by their household. Every second viewer (48 per cent) across mother tongue reported to have taken cable connection for 'variety in television programmes' followed by reasons like 'avoid going to neighbour's place', and 'entertainment or time pass'. 'Others' included varied responses ranging from having taken cable for particular programmes like cricket match or for old grandmother, for daughter researching commercial arts, cartoons for children or women who were at home, for religious channels, for news for father. Other factors included purchase of new TV set, the insistence of children or avoid going to cinema halls. A viewer (Maidservant) from Colaba was given cable connection by her employer in the servant's quarter.

Few viewers reported that they were forced to take cable connection because either it was Cooperative Housing Society decision or because others in the building took it and, therefore, the Doordarshan signals became unclear and the only way to get proper reception was to take the cable connection. About one-fourth (26.5 per cent) of Marathis took cable connection for "not going to neighbour's place" while one-sixth (16 per cent) of Gujaratis reported to have taken cable for 'entertainment/time pass' as shown in Graph 6.2.

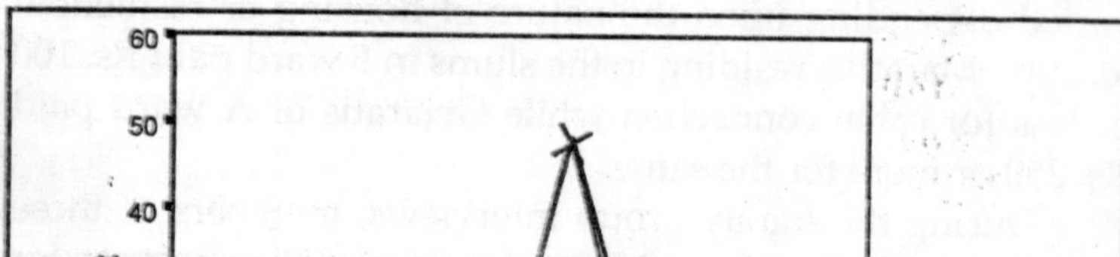

Graph 6.2: Reason For Taking Cable Connection Across Mother tongue

More than one-third (35.8 per cent) of SES-C viewers took cable connection to "avoid going to neighbours" compared to 9.6 per cent of SES-B and 0.8 per cent of SES-A. SES-A opted cable connection for "variety in TV programmes" (58 per cent) and "entertainment and time pass" (13.7 per cent) compared to other SES groups. About six per cent of SES-A and SES-B viewers reported to have taken cable connection for "education/information" compared to only 2.2 per cent of SES-C households.

Family group interviews revealed that the households had adopted for cable connection on an average five years ago but the variation ranged from six months to ten years. All most all the households having cable connection for six to ten years took it for new films being shown illegally on cable networks at that time. Many families including Patankars and Agarkars opted for cable for live telecast of cricket matches on sports channels. Some families expressed that they took cable connection because it came to their building.

Mulays had adopted to cable connection six months back, the reason for taking connection was the daughter who had come to the maternal family for delivery from Mascat and since she was in the house whole day cable was taken for her entertainment. Mehtas also took it for the wife who was at home whole day.

Patels and Kadams expressed that their Doordarshan transmission had become almost invisible and antenna was not working so they were forced to take cable connection at least to make sense of their television set. Kadakiyas also had similar reason besides the father commented that he did not want the daughters to go to the neighbours place. Nisars, Vyass, Tripads also took cable connection to avoid children visiting neighbours place. Daulatjadas children finished Board exams and so important landmark in children studies was over, therefore, cable connection was taken. Tripads with a family size of 12 members expressed that watching films on television is very cheap. Velunkars were music lovers and the grandmother stated that to go to music concerts would be very expensive but television will give that pleasure and would be cheaper.

Modys, Kadakiyas, Lagads, Kadams, Patankars expressed that the cable connection provides entertainment and is good mainly for one-or-two members in the family who were homebound. Kadakiyas and Patankars having working women felt that after work cable is good entertainment and one talks about TV programmes in office also. Baligas, Patels, Patankars expressed that they took cable due to social pressures, feeling that others talked about serials they did not know. It is interesting because on one hand cable was taken to avoid going to neighbours while on the other to talk about TV with neighbours.

Disconnect Cable Connection

Viewers were asked if they 'ever disconnect cable connection at any point of time in a year', so as to know their 'cable dependency behaviour'. Almost nine out of ten (86.8 per cent) viewers had not disconnected cable connection ever since they had taken it. About ten per cent of the households had disconnected cable connection over a period of one year mainly due to examinations of children. 'Others' included responses like thinking about discontinuing cable connection and one Gujarati female viewer stated that their connection was discontinued 'because money was not paid to the cable operator'.

In the course of data collection, researcher visited one household in G ward for family group interview whose neighbour had responded to the individual interview. He remarked that he had discontinued the cable connection because the daughter had come to twelfth standard. Another example was of a working Gujarati male in C ward who was interviewed on Sunday when the cable operator in the area had gone on strike for the hike in subscription money. The man expressed that due to temporary discontinuation of cable he feels helpless and does not know what to do with his time.

Family Television Viewing Behaviour

Television Viewing Time

Viewers were asked to state the time of the day when usually television was viewed in their family. Across mother tongue eight out of ten (78.7 per cent) households watched television at night. Close to half of the households (45.2 per cent) watched television in the evening and proportion of those watching in the afternoon was double than the ones watching in the morning s shown in Graph 6.3.

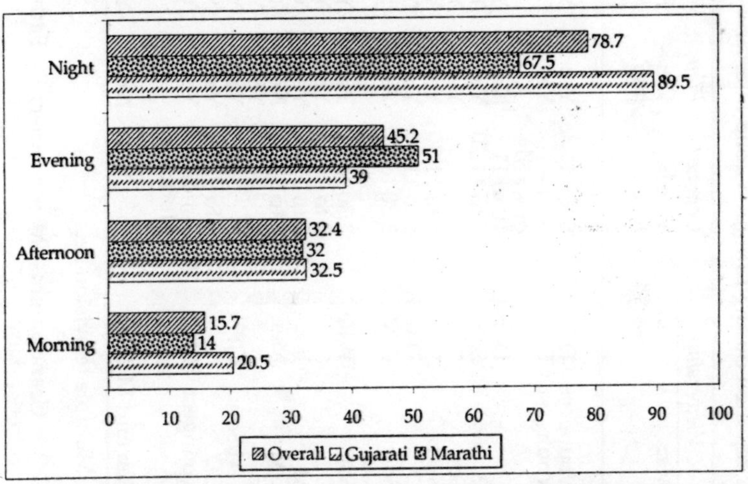

Graph 6.3: Family Time of Watching Television Across Mother tongue

Half of the Marathi households watched TV in the evening (51 per cent) compared to 39 per cent Gujarati households. Higher proportion of Gujaratis watched TV at night (89.5 per cent) compared to 67.5 per cent Marathis. One fifth (20.5 per cent) of the Gujarati households watched television in the morning compared to 14 per cent Marathis. Proportion of Gujaratis watching TV in the afternoon was almost equal to that of Marathis. Analysis also suggested that occupation of Gujarati males impacted their viewing as being in business they had time at home compared to job-bound Marathis who returned home in the evenings and watched television. This viewing pattern was confirmed even by the family group interviews.

Half of the households watched television 'one time in a day' (49.3 per cent) across mother tongue most often at night followed by twice a day, which most often was late evenings and night or afternoon and night. Few (6.5 per cent) households

Table 6.2: Television Viewing behaviour reported during Families Group Interviews

Family Name	*Control of Remote Button*				*Viewership*		*Common*	*Plan*	*TV as*	*Discuss*
	Morning	*After-noon*	*Evening*	*Night*	*Max.*	*Min.*	*Viewing time*	*View-ing?*	*Daily Rou-tines?*	*View-ing?*
Gujarati Households										
Vyas	GM	GM&C	GF	Sons	Youngest	D-i-L	8 to 11	Yes	Yes	No
Kadakiya	×	M&D	F	Youngest D	Youngest D	F	8.30 onwards	Yes	Yes	No
# Nisar	C	GP	×	GP	C	D-in-L	10 to 11	No	Yes	Yes
Patels	Sons	P	P	Sons	P	Sons	7 to 11.30	Yes	Yes	Yes
Ashar	M	Son	Son	F/Son	Son	*	*	Yes	Yes	No
Mody	M	S-i-L	S-i-L	Family	S-i-L	*	7 to 11.30	Yes	Yes	Yes
# Daultjada	×	×	D	Son	D	D-in-L	8 to 11.30	Yes	No	Yes
Vashi	×	C	F	Family	Son	G.M.	8.30 to 11	Yes	Yes	Yes
Marathi Households										
Kadam	×	×	Son	M	Son	Wife	8.30 to 11	Yes	Yes	No
# Agarkar	×	D	Wife	D	D	*	8 to 11	Yes	Yes	Yes
Mulay	×	×	Sons	Family	M	F	7 to 11	Yes	Yes	Yes
Tripad	GM	GM	C	GM	GM	Men	7 to 11	No	Yes	Yes
Lagad	×	×	Wife&D	F	F	*	*	Yes	Yes	Yes
# Velunkar	GM/Son	GM	GM	GP	GM/Son	GF	*	Yes	Yes	Yes
Patankar	×	GP	GD	Family	GD	*	8 to 9.30	Yes	No	Yes
Baliga	×	×	D	P	Youngest D	F	Sunday Morning	Yes	Yes	Yes
Inter-community Household										
Mehta	×	Son&M	×	F	Son and M	F	*	No	Yes	No

× Television was Not Viewed # Multiple television Sets
* No fixed time to watch Television\Not Applicable
Viewers: GM = Grand Mother, GF = Grand Father, G/P = Grandparents, M = Mother, F= Father, C=Children, D = Daughter, D-i-L= Daughter-in-Laws, S-i-L= Sister-in-Law

viewed television 'whole day' as viewing took place by one or the other member of the house. Proportion of Gujarati households watching television whole day was double than that of Marathi households.

In case of Mehtas, Kadakiyas, Modys, Velunkars and Patankars, television was viewed even in lunchtime mostly by members who were at home (grandparents, mother) or those who return home (working daughters/lady coming for lunch at home, son coming back from college). Most male children reported to be coming back from school or tuitions or college switched on the television set for 'relaxation'. Tripads and Patels had municipal water timings in the morning and women of the house switched on the television (earlier they used to listen to radio as commented by them) as wake-up call for others sleeping in the one-room house.

Mr. Lagad sleep late watching TV and wake up late but since he had his own business, it was O.K. as reported by his wife. Similar case was reported by Daulatjadas mother about the unmarried daughter who watches television till late and wakes up late. She commented that earlier she was working, therefore, she used to get up in spite of late night television, now because she is at home; she has no fix time and keeps on watching TV till late night.

Household Space and Television Set

Family group interviews revealed that the physical household space did influence the act of television viewing but families with smaller houses had evolved varied strategies like use of headphones or restrictions on television time.

The viewing behaviour across households and mother tongue varied dramatically. Mulays and Vyass both stay in one-room household kitchen. Mulays head of the family had a rule that television set has to be switched off at 11.00 p.m. even if the programme (most often film) is incomplete. Unlike that Vyass grandfather kept TV set on till he sleeps and the wife commented that many times she gets up in the middle of the night and switches it off. He felt that he gets sleep faster if the TV is on. In the same household, the young daughter view TV till late night using headphones and change the direction of the set to avoid light disturbance to people sleeping in the room. Even in case of Vashis having one-room and kitchen father and son watch sports by keeping television volume on mute while others sleep in the same room. In Patels one room household the working sons use earplugs since last two years to avoid sound disturbance to the sleeping parents.

Mother in the Baligas family (residing in Apartment) expressed concern that the young daughter was watching television alone at night, which she had discontinued, as 'it was not good for her'. Tripads in one room house again had problem with late night viewing. Since the three boys go to early morning (English Medium) school late night television was not allowed on weekdays. Patankars as working couple had early day on weekdays and, therefore, TV was switched off by 9:30 pm.

Decision-making about Television Viewing

Multiple channel choices, remote control facility and preferences for specific target audience programming, may lead to conflicts in television viewing in the household. The viewers were asked about television viewing behaviour in the household through statements, which they were to respond in true or false.

Role of the Eldest of the Family

In response to the statement, 'the eldest in the family decides what will be viewed on TV', seven out of ten (69.8 per cent) households negated the role of eldest. The eldest of the house had 'no say' in deciding about television programmes. Four out of ten (38 per cent) Marathi households had eldest member deciding about the television programmes to be viewed by the family compared to one-fifth (22.5 per cent) of the Gujarati households. Responses of both the sexes remained equal among Marathis while one-fourth (26 per cent) of Gujarati females acknowledged the role of the eldest compared to one-fifth (19 per cent) of Gujarati males.

Family group interviews revealed that family composition, status of the head of the family, relationship equation of the family members all were crucial factors contributing in decision-making related to television-viewing. Table 6.2 shows that the remote control was handled by different members of the households at different times of the day depending upon the family daily routines. The table also shows that most often eldest or youngest member of the household controlled television.

Role of Television in Family Fights

The analysis of responses to the question, 'television is the reason of fights as everyone wants to watch programme of their choice'. Six out of ten viewers (61.3 per cent) across mother tongues witnessed conflicts and arguments due to diverse viewing priorities of family members. Both males and females accepted that TV had created conflict in the family but proportion of Marathi females and Gujarati males was slightly higher. Across mother tongue television had created conflict in the family as evident from the responses. As discussed earlier many households opted to buy additional television set to resolve the conflict.

Family group interviews revealed that cricket matches were most often mentioned as conflict item when viewing priorities were discussed. The common comment was 'no one is allowed to watch anything else if there is a cricket match'. Tripads, Mulays, Daulatjadas and Vyass reported that whenever, mostly Cricket or Football matches are there others in the family are not allowed to watch other things on television. Contrary to that Velunkars stated that they had stopped viewing matches after match fixing scandal. Many individual viewers also had similar comment. Modys household with all women (as men had been to US for work or studies) expressed that cricket matches used to reduce their television viewing out now they decide amicably what will be watched.

The conflicting viewing priorities and tensions, especially in the household owning single TV set, was evident in the nature of discussions that took place in the presence of the researcher. As there were tensions in the family about 'when' television was viewed and 'what' was to be viewed by 'whom', especially children and youngsters.

Ashars young boy of 10 years commented that he does not understand why his mother does not want him to watch AXN, Star Movies etc. He resented his mother's comment that he was mainly viewing cartoons. In case of Mehtas and Vashis the boys were addicted to WWF fights. Vashis head of the family stated that he was happy that the channel had closed down (ESPN had discontinued WWF for a time being) the son immediately commented that it was being telecast on 'Ten sports' (new channel). The college-going boy. at Mehtas responding to parents' disagreement about him watching WWF said, "It may be useful some day and anyway the channel stated that the fights are not real but for entertainment". At Vashis household, mother control the time slots for adolescent boy and girl who had their quota of television viewing. In case of adult families, members usually mutually agreed to common programmes or defined slots. Women who were at home usually view their daily serial viewing quota in the afternoon to avoid conflicts in the evening and night.

In case of confrontations between elders and younger age groups most cases elders compromise. In Patankars household the young girl (7 years) had her TV viewing quota and the grandparents compromise by viewing their quota of TV when she goes to school. Even in case of Daulatjadas the retired father compromised his news viewing to unmarried daughter's demand for serials or films. But in case of Baligas household the father had the control over remote and daughters had to compromise, as was the case at Lagads. In case of Patels (sons do not like serials and parents did not like English films/ programmes) news were viewed as compromising content.

Velunkars, 21 years old grandson watched whatever he likes while the mother and the grandparents mutually agree for Marathi serial or music programmes. In case of Nisars the grand-parents preferred Gujarati programmes while others in the family had different choices. At the time of interview, new television set was bought and will be unpacked in the bedroom where the father commented, "they can watch what they like on it and leave us with what we want".

Time Table for Television Viewing

During individual survey, viewers were asked statement, 'In our family we have decided who will watch which programme, when and we follow that timetable', as to examine coping mechanism for TV viewing conflicts in the household. Irrespective of mother tongue, seven out of ten households (74.3 per cent) television viewing was not a planned activity as they negated the statement. Only one-fourth households planned timetable for TV viewing. Three-fifth (61.5 per cent) of Marathi females and two-fifth (38.5 per cent) of Gujarati females agreed to the statement suggesting that the timetable was decided often to avoid conflicts.

Analysis of television viewing as planned activity indicated that one-third each of the households to have timetable for television viewing were families having 'only adults' and 'adults and children' in the household. Both required planning television time, first to avoid conflicts and second to have discipline, as expressed by viewers during individual interviews.

Family group interviews revealed contradictory finding. All 14 except three households television viewing was planned activity defined by clock. TV viewing occurred most often late evening to night and mostly associated with dinner. As mother of the Vashis household reported, she used to miss her quota of television so decided to serve dinner in front of television recently. Modys reported that morning brunch, cutting of vegetables, stitching or writing letters all take place in front of television set. Irrespective of the mother tongue, women in the Patels, Vyass, Mulays, Lagads, Baligas, Tripads and Patankars households reported of doing all sorts of household work in front of TV set. Most often women reported to 'listened to' the television and watched when sound caught their attention. Many children stated that they did their homework or eat food in front of TV set. In the Ashars household Mr. Ashar keep the audio on mute and read newspaper or at times when the child does craft in front of TV or mother teaches the child he puts headphone and watches news.

The households with multiple sets like Nisars, Daulatjadas as well as Agarkars and Velunkars followed distinct patterns of television viewing. In case of Daulatjadas the old television set has been kept in storage space "to avoid division of family". Father expressed that if he will allow other television set to be kept in son's bedroom they would not come out only and sit with them as they do now because of one TV. But he stated that he and his wife sleep early to allow the son and daughter-in-law to watch what they like. Even among the elderly Velunkars grandfather preferred to watch news and grandmother liked music and Marathi serials. But as commented most often grandmother used to succeed in her programme choices. Daulatjadas and Agarkars reported heavy surfing of channels while watching TV.

Visuals shows different activities carried out by television viewers like Daulatjadas play with grand-son, Tripads women did housework, Patankars and Vyas granddaughter answered phone and played on swing, Vyass and Velunkars watching TV together, Kadam's son watching cartoon, Nisars grandson eat lunch, Agarkars work on computer and relax while watching television.

Television Programmes and Surfing

Two statements with reference to advertisements and sex-violence were asked to examine viewing behaviour in dealing with television content in the household.

Surfing at the Time of Advertisements

'We change channels when advertisements come on television'; this statement was asked so as to ascertain advertisement-viewing behaviour. Three-fourth of the viewers (73.3 per cent) agreed to this statement, meaning they did not watch advertisement but surfed channels when advertisements appeared during the programme. One-third of Marathis watched advertisements (33.5 per cent) compared to one-fifth

VISUAL-4: Activities While Watching TV

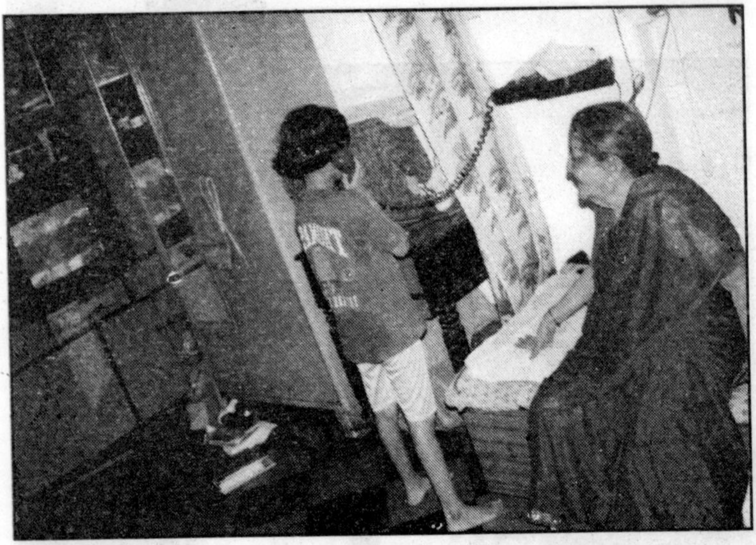

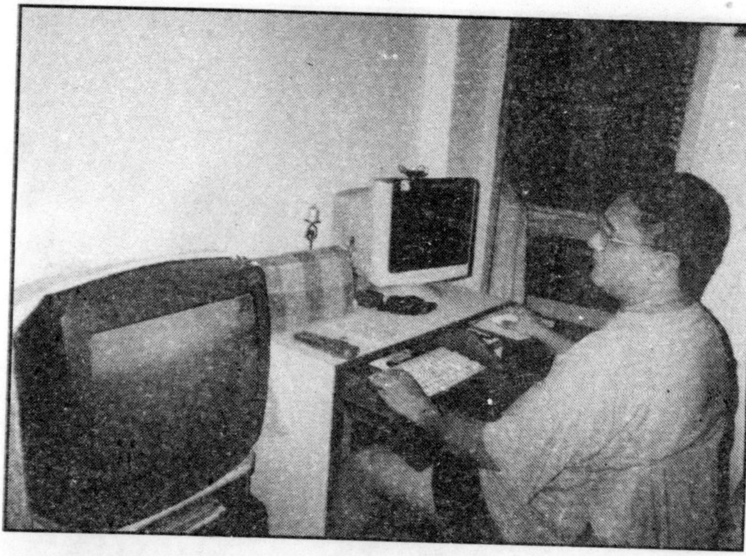

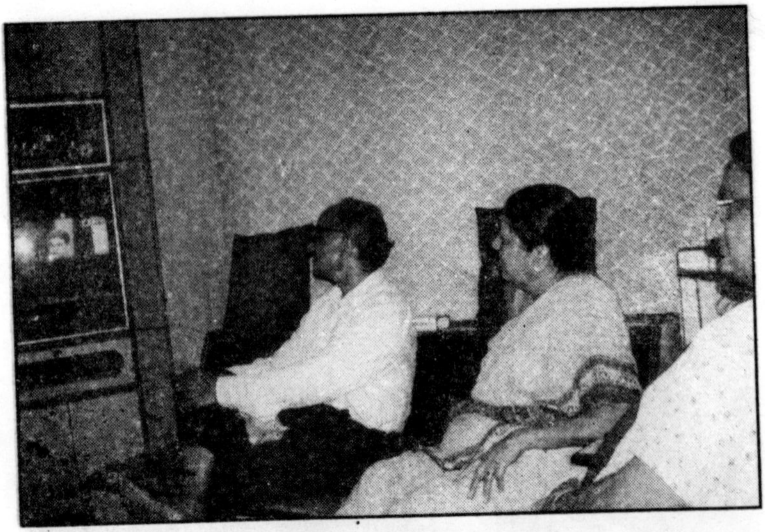

of Gujaratis (20 per cent). As discussed earlier 15 per cent of the Marathi households had television set without remote control. But the analysis indicated that one-fourth of (24.6 per cent) of the households having remote control facilities also watched advertisements and half of those having non-remote set also surfed advertisements by physically getting up and changing channels.

Advertisement viewing behaviour of the households in relation to SES revealed that 16 per cent of the SES-A viewers compared to 35.8 per cent of SES-C watched advertisements. SES and advertisement viewing had inverse relationship.

Sex-Violence on Television

In response to the statement, 'We as a family have got used to watching scenes of sex and violence on TV without changing channels', majority of the viewers disagreed to the statement indicating that viewers were not yet used to watch sex-violence on TV. In spite of being closed-end question, many viewers reacted with additional comments. Viewers who agreed to the statement during interview were 'not happy about the situation' and expressed their inability in watching whatever came on television at that time.

Viewers, most often females, stated that they surf channels, mute the volume or leave the room to avoid embarrassment caused due to certain scenes on television. Most viewers commented that on Doordarshan such scenes were censored but with satellite television channels the number of such programmes has increased. Acceptance of the situation seemed high among Gujaratis compared to Marathis. Analysis of age and acceptability to sex and violence on television indicated that only one-fourth of (26.5 per cent) 40 to 50 years old had got used to watch such scenes on television compared to other age groups.

In Brief

Research showed that the common feature of television ownership was 21-inch, colour television set with remote control facility. Close to half of the household had 14-inch TV set due to smaller sizes of the rooms or one room household. One-fifth of the total households owned multiple television sets. The ratio of Gujarati households with multiple television sets compared to Marathi households was 2:1. In spite of differences in television ownership in terms of type, size and number of sets, there was no difference in the channel capacity of the television sets across mother tongue.

Gujarati households had opted for cable connection earlier than Marathi households. Six out of ten Gujarati households had cable connection for 'more than eight years' compared to four out of ten Marathi households having taken cable in last four years. The years of cable connection ranged from six months to ten years.

The households spent monthly average of Rs. 124 for cable connection with higher deviation among Gujaratis. Gujaratis spent up to Rs. 500 compared to Rs. 250 spent by Marathis. Variety in television programmes was the most often reported reason for taking cable connection. Marathi households had taken connection 'to avoid going to neighbour's house' while Gujarati households opted for cable for 'entertainment and time pass'. There was variety of other reasons including compulsion to take cable connection because of cooperative society's decision or poor signal quality or insistence of members of the family.

Nine out of ten households had never disconnected cable ever since they took it. The only reason for the limited number of household who disconnect cable connection was the examinations of children. Family television viewing behaviour indicated that television was viewed at night. Proportion of household watching television whole day was 2:1 among Gujaratis and Marathis. Most often Marathi households watched television in the evening while Gujarati households watched it at night.

Television viewing decision-making indicates reducing role of eldest of the household. Only one-third of the eldest of the households decided about what would be viewed on TV of which proportion of Marathi households was almost double than Gujarati households. Six out of ten households irrespective of their mother tongue reported to have conflicts due to varied viewing priorities. Only one-fourth of the households had created time tables of viewing in the family to reduce conflicts. Surfing channels was inevitable part of advertisement viewing behaviour but higher proportion of Marathis watched advertisements compared to Gujaratis. Viewers were not comfortable watching scenes of sex and violence on television as seven out of ten of them disagreed to the statement that they have got used to watch such scenes as a family.

Family group interviews confirmed the trends except that most families though did not plan timetables, viewed television by clock, mostly at night. Number of activities were undertaken while watching television. In most families youngest or eldest members of the households controlled the remote control and so the viewing content.

7

Television Viewing Behaviour

As discussed in the previous chapter, television viewing in the household was a ritual characterized by variety of other activities. This chapter elaborates individual television (TV) viewing behaviour in terms of frequency of viewing TV, reason for watching, viewing company, time of watching and amount of time spent, programmes, channel as well as language preferences.

Frequency of TV Viewing

Individual survey revealed that TV was viewed for a daily average of 210 minutes 'almost everyday'. Nine out of ten (91.3 per cent) viewers watched TV everyday. Those reported to be viewing TV five-six days in a week (3.5 per cent) stated that they did not watch TV on Sunday. Except for four Gujarati females (1.0 per cent) everyone else watched TV at least once a week. Females watched TV more regularly compared to males as 94.5 per cent of females watched TV 'everyday' compared to 88 per cent males. Across mother tongue Marathis watched TV more frequently compared to is Gujaratis.

TV Viewing Company

Across mother tongue TV viewing was invariably family activity for more than half (52.5 per cent) of the viewers. Close to one-third (30.3 per cent) of the viewers stated that they did not have 'defined company' as 'whosoever at home' or 'whosoever likes to view TV' watched TV together. About ten per cent viewers reported to be watching TV either alone or with their spouses. Analysis of viewing company with family composition indicates that six out of ten families watching TV with family members were having children or children and adolescents compared to others. Those watching TV with 'whosoever at home', one-third were families with only adults or adults and adolescents.

Compared to other family compositions, almost equal proportion of families of 'only adults' (11 per cent) and adults and children (9.8 per cent) watched TV with spouse indicating that TV viewing takes place without children. About 13 per cent of the families, having children below 12 years had 'alone' TV viewing pattern compared to other types of families.

During family group interviews, parents at Ashars and Agarkars households expressed concern for their children watching television alone as the parents were either working or were out of the house. Ashars reported that they tell their son that they trust him and he should be cautious with what he watches. They reported that they even attempted to lock few channels in the past. Agarkars bedroom set was mostly used by the wife to watch programmes of her choice (Marathi and English) while the husband watches news in the drawing room. But the young daughter was not allowed to watch television alone in the bedroom. Velunkars second set in the bedroom was used by the grandson who watches television alone mostly watch English programmes as commented by the mother. The widowed mother with in-laws in the drawing room watches Marathi programmes.

Reason for TV Viewing

Viewers were given three choices about the reason for viewing television: 'they watch television because: TV is on, watch specific programme or TV is a habit'. Viewers had to choose most relevant reason for watching TV. The choices indicated viewing behaviour in form of: incidental viewing (watch because TV set is "on"), appointment viewing (watch specific programmes), and compulsive viewing (TV viewing as a habit). More than half (58.5 per cent) of the viewers watched TV for 'specific programmes'. About 28.8 per cent of the viewers irrespective of their mother tongue, watched TV as a matter of 'habit' and 12.8 per cent of the viewers watched TV because 'it was on' in the house and was being watched by others in the house. Across mother tongue one-fifth (20 per cent) of Gujaratis did incidental viewing compared to only five per cent of Marathis. Seven out of ten (67 per cent) Marathi viewers watched TV for 'specific programmes' compared to half (50 per cent) of the Gujarati viewers.

When examined in relation to age and reason for watching TV, more than one-third (37.4 per cent) of Gujarati in the age group of 20 to 30 years were found to be watching TV because of habit compared to only 15.1 per cent of 40 to 50 years olds. 28.3 per cent of Gujaratis and 5.5 per cent of Marathis in the age group of 40 to 50 years watched TV because it is on. This may also be due to fewer Gujarati head of the households decide about what will be viewed on TV as observed earlier.

Relatively 'appointment viewing' was higher among females than males as 64 per cent of the females watched TV for specific programme compared to 53 per cent of males. On the other hand, habitual or compulsive viewing was higher among males as 34 per cent of males watch TV because of habit compared to 23.5 per cent females. Equal proportion of viewers across sex did incidental viewing because TV set is "on".

Time Spent on TV

The viewers were asked to self-report time of the day and amount of time spent each time on weekdays (Monday to Friday) and weekends (Saturday and Sunday) for their individual viewing. The analysis was done in terms of time slot, frequency, amount and average time spent with TV.

Time of Watching TV

Prior to analysing the amount of time spent on TV, individual responses were consolidated in terms of 'number of times' TV was viewed in a day. Viewers were found to be watching TV more often than reported with reference to their family TV viewing time. There is a clear increase in number of TV viewing 'times' on weekends compared to weekdays across mother tongue. About 5.5 per cent of the total viewers reported of not watching TV on Sunday. Close to half (47.3 per cent) of the viewers watching TV 'one time of the day' on weekdays reduced to 37.1 per cent on weekends and those watching 'three times a day' increase from 9.5 per cent to 15.8 per cent on weekends.

On weekends proportion of viewers watching TV in the morning and afternoon increases and those watching at night reduces. Many viewers reported that on weekends they watched TV during the day because of the holiday and do not watch TV at night as they go out in the evening. Besides few commented that their favourite serials are not aired on weekends and there is nothing much to watch on TV.

Pattern of TV viewing on weekends was similar across mother tongue except that higher proportions of Marathis watched TV 'whole day' (8.5 per cent) compared to Gujaratis (2.5 per cent). On weekdays, more than half of the Gujaratis (55 per cent) watched TV 'once a day' which reduces to only 37.5 per cent on weekends unlike Marathis where the proportions are more or less similar. Table 7.1 indicates TV viewing time of the day across mother tongue.

Proportion of viewers watching TV at night goes down from 95.8 per cent on weekdays to 70.5 per cent on weekends. One-fifth of the Gujarati viewers (20.5 per cent) and one-third of Marathi (33.5 per cent) viewers watch TV in the afternoon even on weekdays. Almost 40 per cent of the Marathi viewers compared to 24.5 per cent of Gujarati viewers watched TV in the evenings as discussed earlier. A large majority of the Marathis being in service sleep early so do not watch late night TV compared to 28 per cent of Gujaratis who reported to be watching TV after 11 p.m. even on weekdays compared to 15.5 per cent Marathis.

Most frequently watched time slots for morning and afternoon varied on weekdays and weekends. TV was mostly viewed on weekdays in the morning from 6 to 9 a.m., which would shift to 9 a.m. to 12 p.m. on weekends and afternoon slot of weekdays from either 12 to 2 p.m. or 2 to 5 p.m. would shift to 2 to 5 p.m. Evening and night slots were most frequently after 7 p.m. and 9 to 11 p.m. irrespective of the day of the week. Most often women watched TV in the afternoon and night while males watched it in the mornings or evenings and night.

Amount of Time Spent on Watching TV

Viewing time was calculated separately for weekdays and weekends. The same was cumulated for daily average time spent on TV as discussed in Annexure C.

Table 7.1: Time Spent Watching TV by Mother tongue

TV Viewing Time of the Day on Weekday	*Weekdays TV Viewing (in minutes)*				*Total*	
	Gujarati (N=200)		*Marathi (N=200)*		*Mean (minutes)*	*SD (minutes)*
	Mean	*SD*	*Mean*	*SD*		
Morning	11.70	30.34	20.10	46.99	15.90	49.05
Afternoon	22.50	50.12	43.35	71.60	32.92	55.72
Evening	14.70	28.05	32.70	46.80	23.70	37.05
Night	131.85	53.28	125.40	58.38	128.63	56.87
Mean (in hours)	3.0		3.7		3.4	
S.D. (in hours)	1.5		1.9		1.8	
TV Viewing Time of the Day on Weekend	*Weekends TV Viewing (in minutes)*				*Total*	
	Gujarati (N=200)		*Marathi (N=200)*		*Mean (minutes)*	*SD (minutes)*
	Mean	*SD*	*Mean*	*SD*		
Morning	48.90	73.75	49.20	80.70	39.73	77.21
Afternoon	52.80	77.51	58.65	76.22	62.60	76.83
Evening	35.55	55.63	38.55	58.19	39.58	96.67
Night	91.20	79.62	102.15	77.09	55.91	78.46
Mean (in hours)	3.8		4.2		3.98	
S.D. (in hours)	2.4		2.7		2.6	

Viewers spent an average of 3.35 hours on weekdays with SD of 1.78 hours and on weekends the average time spent was 3.98 hours with SD of 2.57 hours as shown in Table 7.1. The average time spent in the morning, afternoon and evening on weekdays double upon weekends while the time spent at night get reduced from 128.63 minutes on weekdays to 55.91 minutes on weekends. Across mother tongue, Marathi viewers spent more or less same amount of time on weekdays and weekends unlike Gujaratis. Gujarati viewers spent less time with TV on weekend nights compared to Marathis also because other media habits of Gujaratis suggested that higher proportion of them watch films in cinema halls and access Internet and computers. Marathis on an average spent 30 minutes watching TV on weekday evenings compared to 15 minutes spent by Gujaratis. Marathi watch TV in the evenings unlike night viewing by Gujaratis.

Analysis of time spent on weekdays and weekends with sex indicated that males watched more TV on week ends (4.3 hours/day) compared females (3.7 hours/day). Education and time spent with TV differed on weekdays as well as weekends. There was an inverse relationship between education and time spent with television irrespective of the time of the week.

Daily Average Time Spent Watching TV

As discussed earlier daily average time spent watching TV was cumulated from time spent on weekdays and weekends. One-third (32.3 per cent) of the viewers spent daily average of 'less than 2.5 hours' followed by 40.5 per cent spending about two and half to four hours watching TV everyday.

Graph 7.1 reconfirms that Marathi viewers spend more time watching TV compared to their Gujarati counterparts. More than one-third (37 per cent) of the Marathis spent more than four hours compared to about one-fifth (18.5 per cent) of Gujaratis. Marathis spent daily average of close to four (3.8 hours) compared to three hours (3.2 hours) spent by Gujaratis. Independent sample t-test was employed to examine significance of difference in TV viewing across mother tongue. Daily average amount of time spent on TV by viewers across mother tongue differed significantly.

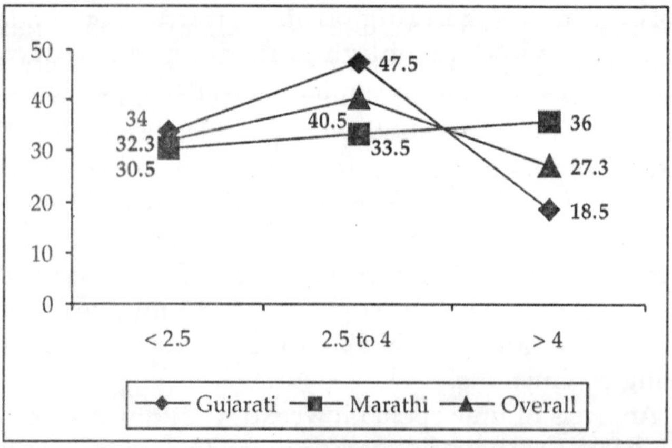

Graph 7.1: Daily Average Time Spent Watching TV Across Mother tongue

Programme Preferences

Viewers were asked about programme type preferences, regularly watched programmes, channel preferences, language preferences and if they watched TV in their mother tongue.

Three separate questions were asked to understand programme viewing behaviour of the viewers. They were asked an open-ended direct question about programmes watched by them regularly. In another question viewers were asked to report in order of preference the type of programmes they usually watch and name the channels on which they watch those programmes. In third question few of the specific channels were named and viewers were asked if they had viewed them in the last week to ascertain their viewing behaviour.

Preference for Programme Types

The interview schedule had listed down few TV programme types along with question on channel preference for that programme type. Viewers were asked to rank the types of programmes watched by them most frequently. Table 5 (Annexure E) shows that viewers preferred diverse types of programmes ranging from information, education, entertainment to enlightenment in form of religious programmes. Eight out of ten viewers (77 per cent) watched more than eight types of programmes and more than half of them (54.5 per cent) had at least four programme type preferences. As first preference, serials were viewed by more than half (52.3 per cent) followed by viewing of news and current affairs by about one-fourth of them (26.8 per cent). Second preference indicated that one-third of the viewers watched films/film-based shows (27.7 per cent) followed by one-fifth (22.3 per cent) viewing news and current affairs and 20.4 per cent watching serials. One-fifth of the viewers preferred music and sports as third preference, which continued to be preference of one-third of viewers as fourth preference. At the fifth preference 42.3 per cent of the viewers reported to be watching music programmes. Marginal viewing of other programme types like cartoons, religious programmes, horror and game shows and women's programme was also reported.

Analysis of first three preferences across sex with mother tongue was carried out. Table 6 (Annexure E) shows serials and news and current affairs as first preference irrespective of mother tongue. It needs to be noted that news and current affairs channels were novelty at the time of data collection as few of the channels had just got launched. Higher proportion of Marathis as compared to Gujaratis reported viewing of films/ film-based programmes. Viewing of news and current affairs, sports and music was more or less equal across mother tongue. Majority of the females (72.5 per cent) reported first preference for Serials as compared to only 32 per cent males. Males preferred news and current affairs (39 per cent) and sports (17 per cent) compared to very few females. Even as second preference, one-third of the total females preferred film/film-based shows compared to one-fifth of the males. There was no difference in viewing of game shows, music, and cartoons across sex. Females also reported viewing of religious and women's programme unlike males. Males also reported to be watching horror shows. Interestingly second and third preference viewing for news and current affairs was more or less equal across sex as TV viewing continues to be family activity.

Regularly Watched Programmes

It was observed that when specifically asked about programmes watched regularly, viewers were more specific unlike question about frequently watched type of programmes. Most often the viewing choices were not focused with reference to regularly watched programmes since viewing was family activity and time bound routine.

In response to an open-ended question about regularly watched programmes or channels, viewers reported 65 different combinations of programmes. The responses based on the names of the shows or channels were categorized and quantified as shown in Table 7 (Annexure E). Analysis revealed that 'serials' were the most preferred programme type as more than half of the viewers irrespective of their mother tongue were watching serials regularly. Serials included Hindi family soaps, English serials, game shows, and horror shows. Across mother tongue there were similarities in viewing choices except for films/music and viewing of Marathi channels. About one-fifth of the Marathis watched films and music programmes and ten per cent of Marathis reported to be viewing Marathi channels regularly unlike their Gujarati counterparts.

Earlier, when asked about programmes type preferences, 26.8 per cent viewers reported to be watching "news and current affairs" as a first preference (Table 5—Annexure E). However, relatively fewer reported to be watching them 'regularly' (Table 6—Annexure E). About 18.5 per cent of the Gujarati viewers and 15.5 per cent Marathi viewers did not have specific programme choices as they watched 'whatever they felt like' or 'were forced by others in the house'. Informative programmes in form of women's programmes, Cookery shows, National Geographic or Discovery channels, Quiz, Religious programmes, Yoga and Sports were viewed regularly across mother tongue along with entertainment programmes.

Analysis of regularly watched programmes by sex reconfirmed earlier discussion. Males watched sports and infotainment channels while women watched women's programmes, cookery shows, *yoga* and aerobics. Table 7 (Annexure E) shows that across mother tongue more than half (54 per cent) of the females watched 'serials' compared to about one-third (36 per cent) of the males. More than one-fifth of males (22.5 per cent) compared to one-eighth (12.5 per cent) of females did not have specific programme choices. Higher proportion of males watched TV out of habit unlike females who do 'appointment viewing'.

Analysis of age versus programme preferences suggested that while 17.2 per cent of 30 to 40 years age group and 10.2 per cent of 40 to 50 years old watched news and current affairs programmes only 8.4 per cent of 20 to 30 years age group watched news and current affairs programmes (Table 8, Annexure E). Viewers in 30 to 40 years age group watched more of news and current affairs and informative programmes compared to others. 40 to 50 years age group watched more of serials and informative programming in form of religious programmes. Interestingly type of serials watched across age group varied. 20 to 30 years age group reported to be watching English serials like "Friends", and "Bay Watch", while 30 to 40 years age group watched family soap operas, 40 to 50 years age group watched mythological or comedy. There was an inverse relationship between age and viewing of film/music, English films/serials and 'no specific show' preferences. Relatively more in 20 to 30 years age group watched TV because of habit and watched it for 'time pass' than other age groups. Interestingly most of the Marathis in 20 to 30 years and 40 to 50 years age groups watched Marathis programmes regularly.

Education influenced programme viewing especially viewing of news and current affairs and serials. More than half (53 per cent) of school educated (SSC/HSC and below) viewers reported to be watching serials regularly compared to only one-third (37.3 per cent) of the post-graduates. Similarly, 20.9 per cent of post-graduates stated that they watched news and current affairs regularly compared to only 7.3 per cent of those educated up to SSC/HSC and below. Even proportion of English serial viewing had direct relationship with education.

Language Preferences

The viewers were asked, 'more often which language programmes do you watch on TV'? Table 9 (Annexure E) shows that almost eight (77 per cent) out of ten viewers watched Indian language programmes, which included Hindi as discussed earlier. About 15.5 per cent of the viewers watched English as well as Indian language programmes. Viewing of only English language programmes was marginal (7.5 per cent). Higher proportion of (10.5 per cent) Gujaratis compared (4.5 per cent) to Marathis watched English language programmes. It is so also because Mumbai being city of trade and commerce English is a language of communication but how far this can be generalized to even Mumbai suburban areas and other cities of India is a debatable issue.

Table 10 (Annexure E) lists down the language combinations reported by the viewers suggesting Indian viewers' complexity of language of media consumption. If the data are pooled together, one-third of the viewers watched TV programmes in more than one language. Gujaratis more often reported to be watching TV programmes in Hindi and English while Marathis mostly watched Marathi, Hindi and English language programmes.

Channel Preferences

Viewers were asked three separate questions to ascertain their TV channel preferences. In response to the question about regularly watched programmes/channels, 54 channels were named by viewers from about 125 TV channels available to them at the time of data collection. Viewers named 256 combinations of channels, which they viewed regularly [Annexure D(iii): p. 206]. There has been broad classification of nine categories of channels as shown in Graph 7.2.

Graph 7.2 clearly indicates heavy dependence on Indian/ Indian-origin general entertainment channels. It also shows similarities in channel preferences across mother tongue except for Marathi/Gujarati language channel viewing as 61.5 per cent of Marathis reported to be watching them unlike only nine per cent Gujaratis watching Gujarati language channels. Apart from general entertainment, new and current affairs channels were followed by music and sports channels. About one-third of viewers viewed movie and regional language channels. Preference for other categories like religious and local cable network was about 15 per cent.

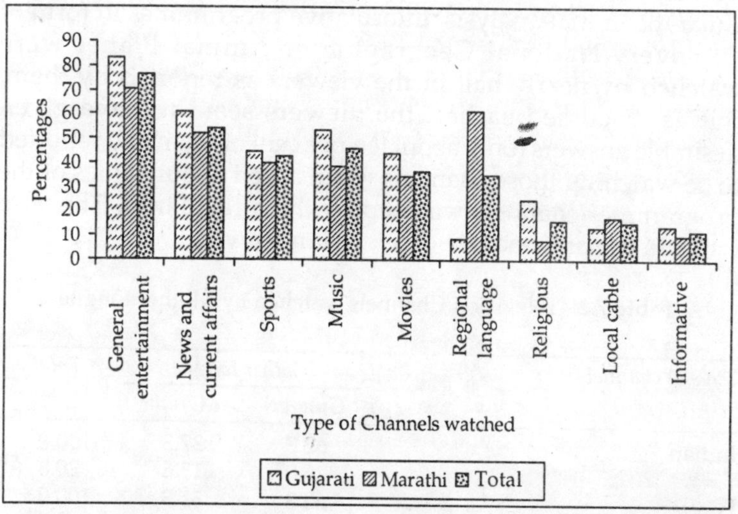

Graph 7.2: Television Channel Preferences Across Mother tongue

Table 7.2 clearly showed that 'only foreign channel' viewing was about four per cent, which was more or less equal to the language preferences discussed earlier. Viewing of 'both' Indian as well as foreign channels mainly indicated viewing of music (MTV or Channel V) or Sports channels or STAR, Sony (General entertainment), CNBC or STAR News (business and news) channels having local content.

In another question, viewers were given the list of channels. The names of the channels were: Doordarshan (Indian), Zee English and Zee MGM (foreign), AXN and HBO (foreign), STAR World and STAR Movies (foreign), MTV and Channel V (foreign), ETV and SNTV (Indian origin), Discovery/ National Geographic and Animal Planet (foreign). Viewers were asked if they had watched (without surfing) any of the channels in the last week.

Interestingly when asked with the specific name of the channel different picture of channel preferences emerged compared to Table 7.2. Close to one-third of Gujaratis had watched only foreign (English language) channels in the last week compared to Marathis and much higher proportion (55.5 per cent) of Marathis had watched Doordarshan compared to Gujaratis in that analysis. Informative programmes in form of Discovery/National Geographic or Animal Planet were watched by nearly half of the viewers as reported by them. But as discussed earlier, the viewers seem to have given desirable answers (only about ten per cent of them had reported to be watching those channels when asked about names of the programmes/channels watched by them regularly). This also indicate dynamic nature of television viewing.

Table 7.2: Television Channels watched by Mother tongue

Type of channels	*Mother tongue*		*Total*
	Gujarati	*Marathi*	
Indian	62.7 26.0	37.3 15.5	100.0 20.8
Foreign	68.8 5.5	31.3 2.5	100.0 4.0
Indian and Foreign	82.1 16.0	17.9 3.5	100.0 9.8
Indian + Local	13.6 4.0	86.4 25.5	100.0 14.8
Indian + Sports/Informative	69.4 25.0	30.6 11.0	100.0 18.0
Sports + Indian and Local	50.0 19.5	50.0 19.5	100.0 19.5
All	15.1 4.0	84.9 22.5	100.0 13.3

Note: List of channels named by viewers in Annexure-E(iii): p. 206.

TV Viewing in Mother tongue

Viewers were asked whether they watched Marathi/Gujarati language channels? Six out of ten viewers (64 per cent) were watching programmes in their mother tongue but there were sharp differences across mother tongue and age. As discussed earlier majority of the Marathis (86.5 per cent) were watching Marathi programmes as against 41.5 per cent Gujaratis who reported to have been watching Gujarati programmes. Further analysis shows that seven out of ten (70.3 per cent) of the Gujarati viewers in 20 to 30 years of age did not watch TV in their mother tongue compared to 86.6 per cent of their Marathi counterparts who watched Marathis programmes.

Many Gujarati viewers commented that the quality of Gujarati TV programmes was not good. One post-graduate Gujarati housewife stated, 'the programmes show primitive culture and I find it difficult to identify with them'. Contrary to that most Marathi viewers including few Gujaratis viewers expressed that Marathi programmes were interesting and well made.

TV Viewing and other Variables

Pearson's correlation coefficient was calculated to examine association of few of the independent variables with three TV related variables: TV viewing frequency, daily average time spent on TV and years of cable connection that made viewing of transnational television possible.

Table 7.3 shows that those who watched television more frequently tend to watch more television per day as both the variables were significantly correlated ($r = 0.137$). Language efficiency of the viewers was directly associated with SES and years of cable connection but inversely associated with daily time spent with TV.

Years of cable connection and SES were directly and significantly correlated ($r = 0.514$) but inversely associated with average daily time spent on TV ($r = -0.104$). SES was found to be inversely correlated with TV viewing frequency ($r = -0.144$) thereby indicating that those from higher SES and better educational status opted for cable connection earlier but tend to watch less television that too less frequently. This suggests that viewers from higher SES, higher educational status and higher language efficiency tend to spend less time with television.

Table 7.3: Correlations of TV viewing behaviour and Independent variables

		Education	*Language Efficiency*	*SES category*	*TV Viewing frequency*	*Years of Cable Connection*	*TV Daily Average*
Education	r	1.000					
	S	.					
Language Efficiency	r	.521**	1.000				
	S	.000	.				
SES category	r	.580**	.436**	1.000			
	S	.000	.000	.			
TV Viewing frequency	r	-.086	.020	-.083	1.000		
	S	.085	.691	.096	.		
Years of Cable Connection	r	.371**	.330**	.504**	-.016	1.000	
	S	.000	.000	.000	.754	.	
TV Daily Average		-.239**	-.155**	-.182**	.137**	-.104**	1.000
		.000	.002	.000	.006	.037	.

[r= Pearson's correlation, S= Signification at 2 tail, * Correlation is significant at the 0.05 level]

In Brief

TV viewing was invariably everyday activity and more than half of the viewers across mother tongue watched TV with their family members. Viewers who watched TV with 'whosoever at home' had mainly adults or adolescents members in the family. More than half of the viewers watched TV to view 'specific programme' but for one-third viewers TV had become habit and for one-fifth it was incidental family activity. Though television was viewed most often at 'night' and/or 'evening' about one-fifth of the viewers watched TV at anytime of the day. Proportion and amount of time spent on TV was invariably higher on weekends than on weekdays except for night time. The analysis of daily average time spent on TV indicated that Marathis watched more TV than Gujaratis.

In terms of programme preferences, there was not much of a difference in viewing of serials, news and current affairs, sports and music programmes. Sex and age differences prevailed in programme-type preferences as well as regularly watched programmes. Females preferred and watched serials and films. Males watched news and current affairs and sports. Proportionately higher proportion of males watched TV as a matter of habit and, therefore, did not have specific preferences in terms of regularly watched programmes. Women did 'appointment viewing'.

20 to 30 years old watched sports, music, films and serials while 30 to 40 years old relied more on information content. 40 to 50 years old watched more serials and religious programmes. SES also influenced programme choices. Viewers with higher SES (SES-A) watched more of films and news and current affairs programmes while SES-C viewers watched more of serials. SES-B viewers did not have specific preferences.

Language preferences of viewers negated cultural imperialism premise and accepted 'active audience' framework as Indian language programmes viewing dominated viewing behaviour. The analysis also revealed contradictions in reporting of television viewing preferences. Responses of question about 'regularly watched programmes' apparently provided more accurate understanding. The viewing of English serials/programmes or channels was not more than ten per cent. Marathis watched TV programmes in their mother tongue unlike Gujaratis. Pearson's correlation analysis of television viewing with independent variables confirmed that higher SES households opted for cable connection earlier but frequency and amount of time spent on television reduces with higher number of years of cable connection.

8

Transnational Television Influences

Boyd-Barrett (1997:119) notes that the country which is affected by media influence either adopts this influence as a deliberate commercial or political strategy, or simply absorbs this influence unreflectively as the result of the contact. This chapter analyses and discusses the viewers' responses related to physical, psychological and social Influences of transnational television on personal self, family and society. Perception of viewers about suitability of transnational television programmes has also been discussed.

Apart from quantitative analysis, qualitative responses collected during family group interviews related to influences of television in general and transnational television in particular, has also been described in three parts: influences on individual self, implications in the family and influences in the social sphere.

Influences on Individual Self

The viewers were asked about influences of transnational television on personal self with reference to six dimensions: physical self (looks, fashion), buying decisions, reading habits in mother tongue, religious activities, thinking patterns and awareness about world affairs.

Graph 8.1 shows that almost nine out of ten (85.5 per cent) viewers considered that television had made them more aware about the world affairs. Close to half (49.5 per cent) of the viewers acknowledged that their personal thinking and buying behaviour were influenced because of television. About four out of ten (38.5 per cent) viewers expressed that their consciousness about physical self in terms of fashion trends and self-image had been influenced by television. However, three-fourth of the viewers felt that their habits related to reading in mother tongue or religion had not changed because of television.

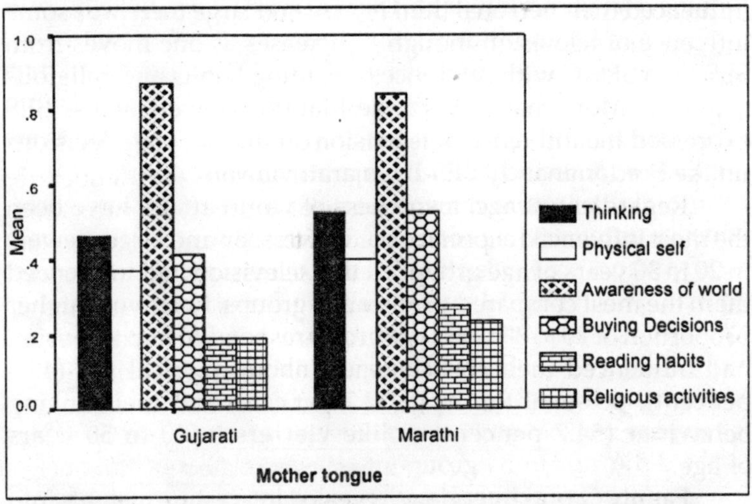

Graph 8.1: Responses About Influence of TTV on Personal Self by Mother tongue

Higher proportion of Marathis expressed influences of television on their religious activities and reading habits. More than one-fourth (27.5 per cent) of Marathis compared to 19 per cent Gujaratis indicated influence of television on reading habits also because higher proportion of them had reported reading habits earlier. During the interviews many of them described change in religious activities by stating that it had increased due to mythological serials and reduced in terms of frequency of visit to the temple.

There were differences in the way viewers across SES and within mother tongue responded to the question of influences on personal self. There is a gradual increase in awareness of world affairs as one moves from SES-C to SES-A irrespective of mother tongue of viewers. On the other hand SES-B seems to be the most influenced with little differences across mother tongue.

Close to half (47.6 per cent) of the Marathi SES-C viewers reported awareness regarding their physical self due to television compared to little less than one-third (31 per cent) of SES-A Marathi viewers and 39.3 per cent of SES-A Gujaratis. More than half (53.8 per cent) of the SES-B Gujarati and majority (63.1 per cent) of SES-C Marathis felt that television had influenced their personal thinking. By and large there was some influence of television though it increases as one moves from SES-A to SES-C with reference to reading habits and religious activities. More than half of the Marathi viewers across SES expressed the influence of television on their buying decisions unlike Predominantly SES-B Gujarati viewers.

Regardless of age, awareness of world affairs have been the most influenced aspect for the viewers. By and large viewers in 20 to 30 years of age group felt that television had influenced them the most compared to other age groups. Relatively higher proportion of 20 to 30 years age group responded that television had influenced their consciousness about physical self (49.3 per cent), personal thinking (52.7 per cent) and their buying behaviour (54.2 per cent) unlike viewers in 40 to 50 years of age.

Family Group Interviews revealed interesting insights into influences of transnational television on an individual self. Vyass family head (grandfather) still performs duties as priest in religious ceremonies. At the time of interview he was watching suspense serial CID. When a character in the serial spoke on mobile phone he commented, "if you have a mobile phone, all your work gets done". During the family group interview, the 18-years old girl in the family commented to her father about the looks of her younger brother (4 years old) wearing a red and blue colour dress. The father did not understand the question. She commented that he looked like 'Spider-man' that comes on the television. During the discussion she expressed that television has much to do with information and fun. When the issue of watching less television at the time of exams was discussed, she very clearly commented, "Television is for mental relaxation, few hours of viewing it even at the time of exams does not make much of a difference" (she failed in her 12th standard examination as reported by her grandparents due to death of her mother. A person in her neighbourhood commented that what she is an outcome of watching fashion on television. She seemed to be addicted to TV.

In Mulays household, the daughter while feeding her four months old daughter kept on saying "after the break" (line often repeated in-between programmes on television) and her brother proudly informed me that the young girl does not eat her food if television set was not switched on.

Influences on Family

During individual survey viewers were asked about the five dimensions of family life, which are generally perceived to be influenced by television. They are: Family Interactions, communication with the Head of the family, frequency to eat ready to cook food, family get-together and social visits. First two being family communication issues, frequency of ready to eat food at home is to do with the acceptance of 'fast food' concept and last two to understand the way family and community had opted to cope with 24-hour transnational television.

Graph 8.2 shows the direction of change across mother tongue. Higher proportion of Marathis reported influences of transnational television on family compared to Gujaratis as is evident from Graph 8.2.

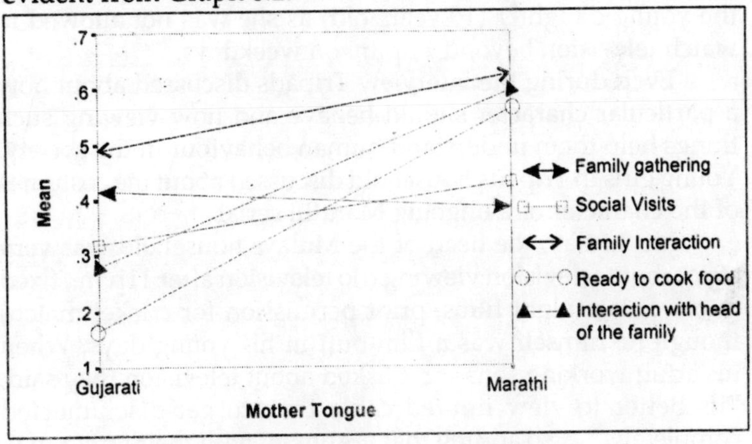

Graph 8.2: Dimensions of Influence of TTV on Family by Mother tongue

Family Interaction

One-third (39.6 per cent) of viewers expressed changes in family interaction patterns due to television viewing across mother tongue. Relatively higher proportion of Marathis expressed increased family interaction while Gujaratis indicated reduced interaction due to TV. Almost one-third viewers of 40 to 50 years age group reported felt reduced family interaction due to the presence of television. In terms of SES, proportionately one-fourth of SES-C viewers expressed increased family interaction compared to 35.9 per cent of SES-A viewers who expressed that family interaction had reduced because of television.

During family group interviews, majority of the families expressed that they discuss about the television programmes as a family. In most cases women discussed the storyline of a missed episode of a soap opera and men about sports matches. Many women commented during family group interviews that neighbours, relatives and friends always discussed about previous night episode of a soap opera. Patels, Daulatjadas, Mulays, Velunkars, Tripads, Agarkars reported that they do discuss among themselves television serials in terms of events and characters and how much it reflects present day social life. Agarkars stated that it was done as an educational process for the young daughter (10 years old) as she was not allowed to watch television beyond 11 p.m. on weekdays.

Even during the interview Tripads discussed about how a particular character should behave and how viewing such things help them understand human behaviour in the society. Young girls in Tripads household discussed about the costumes of the character of a ongoing Marathi serial.

Mr. Mulay, the head of the Mulays household was very strict about television viewing (no television after 11p.m., fixed quota of late night films, prior permission for cricket match) though he himself was a film-buff in his young days. When his adult working sons were asked about television they said, "its better to view limited cable than to get disconnected completely." Also adding that argument with their father may lead to discontinuation of the cable connection. The mother and daughter also commented that there was lot of tension between the father and sons because of television-related restrictions after recently taken cable connection.

Nisars head of family (grandfather) felt that alcohol drinking had increased due to television. His wife felt that television had not done well to anyone. She disliked her husband watching Discovery/Animal Planet because she stated; 'lot of fish was shown'. Both the grandparents felt that the parents have responsibilities towards what children watch on television. The daughter-in-laws in the family expressed that child get tired of television viewing, therefore, there is nothing to worry about television. At the time of interview, the child of 4½ years switched on the television and informed the researcher that the TV set does not have remote control. He changed channel on the converter box and tuned cartoon channel while his mother fed him lunch. Once the lunch was over the child switched off the television and both the brothers went down to play. The elder one also sang the cartoon title track. The family had just bought new TV set to be unpacked in the son's bedroom.

Frequency to Eat Ready-to-Cook Food

Television advertisement promotes packaged food, ready-to-cook food and beverages as lifestyle statements. Nine out of ten (91.5 per cent) of the Gujaratis reported 'no change' in their food habits as against 65.5 per cent Marathis. One-fifth of 30 to 40 years age group reported increase in ready-to-cook items compared to other age groups. About one in ten households across SES reported increase while two in ten SES-C household reported 'decrease' in frequency to eat ready to cook food due to TV. The reason cited mainly by Marathi women was a result of their TV viewing of cookery programmes leading to cooking of food items at home instead of buying them from the market.

Family group interviews revealed a different issue related to food habits in the family. Mrs. Baligas reported that her daughter does not remember the name of the vegetable she ate yesterday because she eats while watching TV and did not even noticed her favourite vegetable. Mrs. Agarkar stated similar situation about her daughter (10 years old) who extends her dinner/lunch to continue watching TV and many times eats much more influencing her health. The girl is obese. Referring to the influences of TV Mrs. Agarkar remarked that advertising has made her daughter and others of her age well informed about 'free offers' and more fashion consciousness. Also demand for advertised food products is more frequent also because TV reminds it everyday.

Family Get-together and Social Visits

Viewers' social interaction with friends and relatives (social visits) and their openness to allow others to visit them for get-together was asked in relation to TV. More (11.5 per cent) Marathis expressed increase in social visits compared to six per cent Gujaratis. Across mother tongue the difference in responses about family get-together is marginal. Those expressing reduction in social visits commented that because of television people prefer to watch television than to visit others.

About one-fourth (23.6 per cent) of SES-A Gujaratis and one-fifth (21.8 per cent) SES-B Marathis felt that television had mostly reduced 'social visits'. About 23.9 per cent of SES-C Marathis reported that occasions of family gatherings had increased because of television. Gujaratis expressed increased family gathering as against increased social visits because of the 'family soaps' on television, which has glorified the values of joint family. Viewers reporting reduction in social visits felt that television hindered social interaction. Many Gujarati females reported that if the telephone rang while they were watching the serials they would not like it. Also they experienced that if they called up others at their television viewing time they did not receive proper response.

Family group interviews gave insights into diversity of experiences around television set and its influences on family. Mulays and Daulatjadas felt that television was responsible for reduced social life.

The grandmother (ex-government employee) of Velunkar household had an apt comment about influence of transnational television on the household. She was very much against television (though she was the main television viewer in the household). She commented, "the big houses, money shown in television appear 'reality' to children. Their approach to money, values and culture has changed because of today's television. Films had limited influence but television has come into the houses. Marathi household now has money because of working parents and parents go out of the way to provide facilities to their children. Television is definitely to be blamed for the situation."

Mehtas are an inter-community household, the Gujarati head of the family watches news, sports and mythology serials. The Marathi mother watches Marathi serials and the son watches WWF, and English Films. Mr. Mehta sarcastically commented that his son watch TV till late night and was not studying. So, the question of his studies getting influenced because of TV does not arise. Mrs. Mehta informed the researcher about distance education courses run by local cable operator related to cooking/catering, which she thought were useful for women who were at home.

Vashis felt that TV was influencing their son's studies but their son did not agree. The mother said, "he winds up his studies to watch television". Mr. Vashi also commented, "it was good that his children were not hooked to films like other families". The family had a relative's son staying with them for his college studies. The boy's parents were working back home and he used to watch lot of television but in Vashis household the father had set viewing restrictions.

Modys household male members had gone to US for studies and work. The only women household had eldest woman (sister-in-law) who was at home felt TV was a blessing. The women felt that TV couldn't influence beyond family environment and whatever was shown on television does exist in some families. She expressed that in certain families people fight over programme choices and even break the TV set, but they did not have such a problem.

In the Daulatjadas family, the father commented that as a boy he was not allowed to watch movies, television was non-existent. However, today his grandson started watching TV at a very young age of three months. According to him the present generation parents encourage the babies to start watching television right away so that they could be free from the responsibility to look after the child. The unmarried girl (for whom a suitor was being searched) stated that serial characters were like her family members and her brother teased her on

Thursday saying, "your three sister-in-laws (characters of a daily television soap opera) have gone to their maternal house till next Monday (because the serials telecast is from Monday to Thursday)". They felt that things on TV should not be taken seriously and it was not worth spending their time watching the serials but it was good for entertainment.

Influences on the Society

When viewers were asked 'if satellite television had influenced society in any way', six out of ten of the viewers responded in affirmation; proportion of Gujaratis (53 per cent) was higher than Marathis (47 per cent) expressing any role of TV in society.

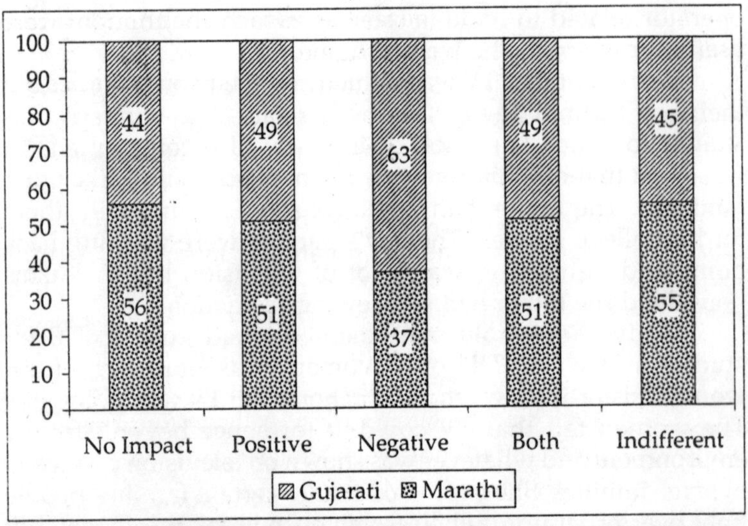

Graph 8.3: Nature of Influence of TTV on Society by Mother tongue

The nature of influences of transnational TV was categorized as 'Positive', 'Negative', 'Both' and 'indifferent'. Negative influence of transnational television was perceived by one-fifth of the viewers whereas equal proportion of them perceived it as 'positive' or 'both' (positive as well as negative). On the whole no clear picture emerged from the analysis of nature of influence but it seems 'denial of influence' by 36 per cent of the viewers. As shown in Graph 8.3 higher proportions of viewers perceiving 'no influence' were Marathis

(56 per cent) compared to Gujaratis (44 per cent) and almost double the Gujaratis perceived it as 'negative' compared to Marathis.

Table 8.1: Responses about Nature of influences of TV on Society by Mother tongue

Nature of Influence	*Mother tongue*	
	Gujarati	*Marathi*
Positive	• Women get to know outside world • World at your doorstep, see things sitting at home • Children learn new things, improve language • Makes people aware and open-minded • People value social relation more	• Increased knowledge about world • Children get knowledge, lose their fears • More choices for entertainment
Negative	• Children know more than us, learn to black-mail us, learn bad things • Violence, western culture and clothing • Make people consumerist • Reduces social activities • Increase in extramarital relations, promote curiosity about male-female relations • Glorifies joint family	• Influences children's studies, reading habits, social communication • Not good for our culture • Make people couch potato • Change in ethics, clothing, life-style • Approves extra-marital relations, promote approach to get things without efforts, quickly and easily • Youngsters and women imitate things • Numbness to violence and vulgarity
Indifference	• It depends on the viewer what to choose for himself/herself • It depends upon which channels you watch • In our family we do not watch certain channels. It depends upon family.	
No Influence	• How can television influence human mind? • It depends upon the individual who watches it and not on the television • If taken personally, it can influence but I do not think it can influence society	

The viewers were asked an open-ended question about what did they perceive was the influence of transnational television on society. Table 8.1 presents qualitative responses with reference to nature of influence reported by individual viewers.

Family Group Interviews reported few experiences related to influences of television on society. Mrs. Mulay felt that "what we try to cover up (hide) gets uncovered (disclosed) by TV and because of TV people do not want to meet each other as they are hooked to their respective serials." She felt that because of television relatives do not come over and most of the subjects on television were love or murder, which was not good for children. She had a clear opinion that TV creates a feeling among children and young girls that 'it is o.k. to be a second wife' (due to adultery and extra-marital issues portrayed in soap operas). She was the most television-addict member in the family but commented that 'television should be thrown out', and felt that she views it because there was nothing else to do. She narrated an incident of a young girl in the neighbourhood. The maidservant had not reported to work in the neighbourhood, and the girl commented that she must have been murdered. The girl's family was watching suspense thrillers regularly.

Daulatjadas parents also commented that television promotes 'status consciousness and importance of possession'. The message propagated are 'spending' and discourage 'saving'. The working son kept on surfing TV channels during the interview. The girl reported that she watches fixed serials and during commercial breaks kept track of English or Hindi films on other channels. Once the serial was over, the viewing shifts to film and so on till she feel tired. When asked if television influences her in anyway she expressed that it would be difficult to know but at the same time has a feeling that television must be influencing her thinking.

Kadakiyas felt that the children were the main victims of television as their building is much quieter now unlike earlier days because children instead of going for outdoor activities watch much more Television. The father felt that even amount of homework has increased influencing free time available to children. Their elder daughter who was also working with children in a voluntary organisation stated that the children's creativity is getting influenced because of Television. The mother felt that there was a change in the viewing behaviour of her two daughters as the elder one viewed television late and had a different taste while younger one watched much more of television. The younger one is mainly interested in English films and serials. Family had a consensus that the family serials were ridiculous and they enlarge and generalize exceptions. Mrs. Kadakiya expressed that TV is making alcohol and extra-marital affairs acceptable in the society. Mr. Kadakiya came to Mumbai in 1937 and had witnessed Chawls getting transformed into blocks and families separating and shifting to suburbs. He felt that while industrial revolution reduced physical work for men, electronic and computer revolution reduced work for human mind.

Mr. Lagad, a self-employed printer, commented that his parents (having five sons) felt that television was bad for women because after watching it they separate from their parents. He had just bought a house in a slum for Rs. three lacs and had his office in the attic of the house. Mrs. Lagad stated that she watches Marathi films and cries with the character and so her daughter of six years. Her husband commented that his wife's film can come some other time but news can't wait. So, when Mr. Lagad is in the house he controls the TV set. He felt, 'television gives you new ideas but one has to choose what one wants in life. One cannot get carried away' but the wife felt, "whatever is happening in the society is shown on television but it has little to do with her life".

Mrs. Baliga took tuitions at home and used to watch television in the afternoon. She stopped TV viewing in the afternoon as her servant used to finish work very quickly to watch TV. She kept on remarking that she does not watch much of television, at the same time commented that a particular character (Tulsi) in the Hindi daily soap teaches good value system. She reported an incidence about how she refused her tailor to visit due to her daily soap timing. But eventually both (she and her tailor) watched the serial together as it was also his favourite serial. She felt that the community culture is changing due to TV.

Kadams head of the family work in Bombay Municipal Corporation as a clerk. Their son is six years old. He was watching cartoon even when the researcher visited them on Sunday morning by keeping the volume on mute. Mr. Kadam stated that they do not watch "certain" programmes because of the child. The child studied in English medium school and his favourite cartoon was Tom and Jerry. The mother like Lagads' wife felt that whatever happens in the society is shown on television.

Grandparents of Patankars expressed that since they do not go out, TV gives them world-view sitting at home. The family having working-couple makes it a point to go to park on Sunday evening to avoid TV. The parents commented that the daughter does demand television advertised products influencing family budget. The working wife reported that she keeps track of "free offers" in advertisements.

Tripads, fifth generation in Mumbai, have been staying in their present one-room house for about 100 years. The grandmother who was the most attentive viewer of TV said, "TV tells you what you should do if you confront similar situation in real life". They felt Hindi serials exaggerate a lot. The grandmother also stated that fifty years ago she was teased by the rowdies on the street so, it is not something that was happening only today. TV is good as it entertains you sitting at home.

Velunkars had lost their son due to tobacco-eating habits as reported by the widowed wife. The family felt that television is all about entertainment and is really good. There are educational messages on television but no one listens to them. Velunkars felt that people do not want serious things and television viewing does not require any work on our end while the people who appear on television gets paid. Agarkars also expressed that no one is interested in the educational aspect of television and for the working couple when everyone at the work place discusses about the previous night episode of the serial, it is a problem.

Interesting difference in the viewing pattern during the year was discussed during group interviews. Mrs. Patel also commented that their sons watch much more TV in monsoon as they cannot go down and hang around with their friends due to the rains. Mrs. Ashar expressed that in monsoons, television comes to her rescue as she prefers her servants sitting and watching TV than going down and meeting other servants which gives them wrong ideas. Many Cuffe Parade viewers having active social life expressed concern about TV viewing by their servants or even children who were left alone at home when they go out. Many remarked that TV exposes servants and children to wrong ideas. Few reported to have locked channels on their servants' TV sets. A resident of affluent locality (Maker Tower, Cuffe Parade) commented during individual interview that her servant now uses products she uses because of TV advertisements and advises her to buy a soap detergent advertised on TV because it is good for her palm.

Suitability of Transnational Television Programmes

During the individual interviews, the viewers were asked if they found any television programmes or channels not suitable for themselves or for their family and if so which were they and why did they feel so. Exactly half of the viewers reacted by saying 'I do not watch such programmes and so I have no idea if they are unsuitable'. About one-third (36 per cent) viewers felt that programmes on television were unsuitable.

Graph 8.4 shows that more than half of the Gujaratis (57.6 per cent) were of the opinion that television programmes/ channels were unsuitable compared to 42.4 per cent Marathis. About 14 per cent of the viewers expressed that there is nothing unsuitable on television as it is a question of how one views things that are shown on television.

Both the groups justified their viewpoints and when asked the names of such unsuitable shows or channels they named from Fashion TV to popular soap opera *'Kyunki Saas bhi Kabhi bahu thi'*.

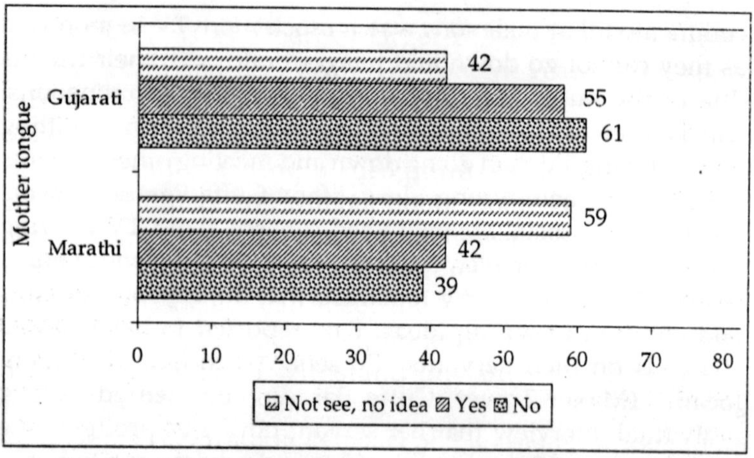

Graph 8.4: Suitability of TTV Programmes by Mother tongue

Table 8.2 lists the type of programmes and names of channel viewers found unsuitable and Table 8.3 shows the reasons cited by viewer for finding these programmes/channels unsuitable. Majority of Gujaratis found English films on HBO, STAR movies, AXN as unsuitable with which Marathi males did not seem to agree as shown in Table 8.2. Close to half of the females felt that music videos on music channels as well as local cable network music jukeboxes were unsuitable. More of Marathi males responded to varied 'other' programmes/ channels.

Table 8.2: Percentage Distribution of Names of Unsuitable Programmes/Channels and Sex by Mother tongue*

Names of Programme/ Channels	*Males*		*Females*		*Total*
	Gujarati (N=43)	*Marathi (N=25)*	*Gujarati (N=40)*	*Marathi (N=36)*	*(N=76)*
FTV	23.3	36.0	27.5	22.2	50.0
English serials/movies	53.5	12.0	42.5	44.4	73.7
MTV/Channel V/My channel/itv	32.6	28.0	47.5	47.2	75.0
Others**	23.3	36.0	25.0	19.4	47.4

* Multiple Response

** Horror serials, Family Drama, Advertisements, Southern channels, cartoons, Animal Planet, STAR Sports.

The reasons for finding the above stated programmes/ channels unsuitable were vulgarity or non-suitability to the Indian culture as shown in Table 8.3. More of Gujarati females compared to Marathi females found shows vulgar and indecent. One-fourth of the viewers irrespective of sex and mother tongue found shows 'unsuitable' to Indian culture and indecent. Most males felt that television taught wrong things to women and children while most females felt that children were getting adversely influenced and, therefore, the programmes were unsuitable for the family.

Table 8.3: Percentage Distribution of Reaction to Unsuitable Programmes/Channels by Sex and Mother tongue*

Reasons for feeling of indecency	*Males*		*Females*		*Total*
	Gujarati (N=43)	*Marathi (N=25)*	*Gujarati (N=40)*	*Marathi (N=36)*	
Vulgarity	41.9	20.0	60.0	38.9	80.3
Violence	6.9	0.0	10.0	5.6	11.8
Unsuitable to our culture	23.3	24.0	25.0	25.0	46.1
Spoils children/women	32.6	20.0	25.0	11.1	43.4
Nothing to learn	0.0	20.0	0.0	11.1	11.8
No particular reason	6.9	8.0	5.0	8.3	13.2

[* Multiple Response]

The factors reported as 'unsuitable to Indian culture' were reported to be extramarital relationships, alcoholism, smoking, sexual permissiveness, obscenity and fights in the family. There was concern about the way children were getting influenced in terms of reduced outdoor activity, violence, imitating television content, addictive to television, watching things not suitable to their age. One Gujarati man commented, "satellite channels show twenty-second century content while we are still in twenty-first century". The unsuitability for Hindi serials was cited mainly because of "why know someone's family problems, boring and repetitive and unimaginative treatment, giving wrong ideas to women to fight for rights" by few viewers.

In Brief

Analysis about influences of transnational television on individual self in six dimensions: physical self (looks, fashion), buying decisions, reading habits in mother tongue, religious activities, thinking patterns and awareness about world affairs suggested that nine out of ten viewers acknowledged "information provider" role of television and felt that television had made them aware about 'world affairs'. There was not much of difference in responses across mother tongue except for reading habits and religious activities. Close to half of the viewers accepted that television influenced their 'personal thinking' and their 'buying for self and others'. Two-fifth of the viewers had become conscious about their personal self because of television. Higher proportion of Marathi viewers compared to Gujaratis expressed that television had influenced their reading in mother tongue.

In terms of age higher proportion of 20 to 30 age group felt that television had influenced them psychologically (personal thinking) and materially (buying, consciousness about physical self) compared to viewers in 40 to 50 years of age group. Proportionately higher proportion of SES-A reported influence in terms of awareness about world affairs; SES-B in buying behaviour and SES-C reported influences on personal thinking and consciousness about physical self.

Family group interviews also indicated that viewers personally realize and accept that television is influencing them but most of them find it difficult to articulate the nature of influences and direction of change.

Influences of television on family included family communication (family interaction and interaction with the head of the family), social interaction (social visits and family get-together) and behavioural change in terms of eating ready-to-cook food at home. Family interaction was found to be most influenced and eating habits was considered to be the least influenced.

Marathi viewers reported increased family interaction because of television unlike Gujaratis by citing reason of spurt of family soap operas leading to revival of joint family values. Higher proportionately of SES-A viewers and 40 to 50 years age group reported reduction in family interaction unlike others. Social interaction in terms of visits to others house and get together of family in general indicated decrease but proportionately more of Marathis reported increased social interaction like family communication unlike Gujaratis.

Six out of ten viewers expressed that television has influenced society in variety of ways. Higher proportion of Gujaratis felt that the influence of television was negative compared to Marathis. Marathis were found to be more tolerant towards television content compared to Gujaratis. Interestingly men felt that TV was spoiling women and women expressed that children were getting influenced by TV.

Family group interviews indicated that families felt that the 'other' family was getting influenced and not their own. Parents of young children were found to be anxious, yet few voiced that television was a learning tool for their children. Viewers irrespective of their age, sex and SES expressed that television had no relation to their own life yet it was a good entertainment. Lower SES (SES-C) viewers especially women felt that television projects 'reality' unlike viewers from upper SES (SES-A).

Transnational television programmes were found to be unsuitable by half of the viewers. Other half took a stand that they had not watched 'such' programmes and, therefore, had 'no idea' if they were unsuitable for themselves or their family. Fashion TV, English films on HBO, STAR movies, AXN, music videos on music channels as well as local cable network music jukeboxes and variety of 'other' programmes/channels were reported as unsuitable. The reasons cited for unsuitability were vulgarity, extramarital relationships, alcoholism, smoking, sexual permissiveness, obscenity, and fights in the family, violence.

9

Transnational Television, Individualism and Consumerism

Present research operationally defined individualism as "Person's strong sense of individual identity as opposed to family, self-assertion and a belief that he/she is primarily responsible for his/her life as against social obligations and interdependence to the family and community". Consumerism was defined as "an urge to acquire and possess material goods, money as criterion for success and high value attached to physical pursuits." The scales were constructed and validated using expert panel as well as sampled viewers as elaborated in Chapter 4.

This chapter presents review of values of individualism and consumerism and examines the findings of the present research with reference to transnational television viewing and these values across Gujarati and Marathi viewers.

Values of Individualism and Consumerism

Values are psychological constructs and in a societal/cultural context they define what is right and what is wrong. What is right and what is wrong is a complex psychological synthesis of ideas about what is generally valued and thus seems to be good to be pursued as a member of a particular culture/society.

Kitayama *et al.* (1993: 4) remarked, "'core cultural ideas' can be seen in 'key ideological and philosophical texts and institutions at the collective level'. These foster cultural shaping of psychological reality thereby affecting 'customs, norms, practices and institutions'".

Ahuja (1993) comments that while Indian nation-state has diversities of races, religion, castes, tribes, languages, social norms, cultural and sub-cultural beliefs, political ideologies, there is a unity in terms of economic, social and political philosophy. And being an individual Indian citizen "has certain responsibilities and duties to perform". These 'responsibilities' include responsibility towards family, community and society. This applies to Asian cultures in general and Indian culture in particular.

Individualism

"Asian values" allegedly stress patriarchy, family deference, community, unanimity, a disciplined and obedient citizenry, and an authoritarian state. According to this view, "Western values like human rights, human dignity, freedom of the press, religion, and speech are alien impositions that have no rightful place in an 'Asian' context" (Keniston, 2002).

"Collectivist cultures emphasize the value of the group or society over the individual, for example, China. In contrast, individualist cultures such as the United States see the individual as the primary entity of value with groups outside of the family assuming secondary importance. The Western emphasis on the individual holds that emotions are important in relationships while Eastern societies are more concerned with roles than any notion of an intrinsic self" (Madden, 2001).

According to Kazdin (2000: 176), "Individualism is a cultural syndrome organized around the importance of individuals in social life. Collectivism is a cultural syndrome organized around importance of collective (family, co-worker, tribe, religious group, political group, philosophical viewpoint, nationality, social class, race)".

Individualism and its polar opposite collectivism, have strong implications for nature of intimate family relationships. Individualist cultures allocate priority to goals concerning the personal identity of individuals, while collectivist cultures emphasize the value of the extended family or the immediate community" (Hui and Triandis, 1986: 225-228).

"Political and social philosophy that places high value on the individual...Individualism stresses self-reliance, privacy and respect for other individuals. Negatively it embodies opposition to tradition, to authority, and to all controls over the individual, especially when they are exercised by the state" (Britanica,1983: 339). "The self is defined as independent and autonomous from collectives. Personal goals are given priority over the goals of collectives. Social behaviour is shaped by attitudes and perceived enjoyable consequences. The perceived profit and loss from a social behaviour is computed, and when a relationship is too costly it is dropped" (Triandis, 1996: 409-410).

"Individualism places the individual at the centre of a framework of values, beliefs and convictions. Increasingly, it is being expressed as self-centredness, the gratification of personal wants, a pre-occupation with entitlements, an abrogation of responsibilities and a withering of collective effort" (Eckersley, 1999).

Oommen (1982: 64) in the context of the "Indian family" ("the term no less confusing academically and popularly" as commented by Chakrabortty, 2002) defined individualism as a value that "fosters the autonomy and freedom of individuals in the decision-making processes in the family-choice of mates, acquisition and management of personal property, establishment of separate household after marriage etc. This is contrary not only to the spirit of familial collectivism ingrained in the traditional value of holism (primacy to the group over the individual, facilitates familial collectivism) but also promotes questioning the legitimacy of the established authority of the elder male, anchored in the principle of hierarchy".

Hofstede discussed individualism *vis-à-vis* collectivism in the work situation from organisational perspective. He also developed 'low individualism index' and 'high individualism index' as shown in Table 9.1.

Table 9.1: The Societal Norms about Individualism

Low Individualism Index	*High Individualism Index*
• In society, people are born into extended families or clans, which protect them in exchange for loyalty.	• In society, everyone is supposed to take care of him or herself and his or her immediate family.
• 'We' consciousness.	• 'I' consciousness.
• Collectivity-orientation.	• Self-orientation.
• Identity is based in the social system.	• Identity is based in the individual.
• Emotional dependence of individual on organisations and institutions.	• Emotional independence of individual from organisations and institutions.
• Emphasis on belonging to organisation; membership ideal.	• Emphasis on individual initiatives and achievements; leadership ideal.
• Private life is invaded by organisations and clans to which one belongs; opinions are predetermined.	• Everyone has a right on a private life and opinion.
• Expertise, order, duty, security provided by organisation or clan.	• Autonomy, variety, pleasure, individual financial security.
• Friendships predetermined by stable social relationships; but need for prestige within theses relationships.	• Need for specific friendships.
• Belief in group decision.	• Belief in individual decisions.
• Value standards differ for in-groups and outgroups; particularism.	• Value standards should apply to all; universalism.

Source: Hofstede (1984:171).

Consumerism

Consumerism has two connotations. The first in terms of protection of the interest of the consumers and the second refers to a high rate of consumption...in fact the first meaning has almost been given up...Consumerism which refers to increased consumption of different goods and the promotion of more and more outlets for acquiring theses goods in one of the most significant developments of the last two decades...the message is beamed through media (through the advertisements urging people to buy, buy, and buy) (Narasimhan, 1998: 34).

"Consumerism is the orientation of individuals towards purchasing products and services"(Singhal and Rogers, 2001:126). "According to the thirteenth century theologian, St Thomas Aquinas, the seven deadly sins are pride (self-centredness), envy, avarice (greed), wrath (violence), gluttony, sloth (apathy) and lust; the seven cardinal virtues are faith, hope, charity (compassion), prudence (good sense), temperance (moderation), fortitude (courage, perseverance) and religion. Consumerism effectively reverses these lists" (Eckersley, 1999).

Bearden *et al.*, (1993) have noted many definitions of consumerism. "The idea that goods are a means to happiness; that satisfaction in life is not achieved by religious contemplation or social interaction, or a simple life, but by possession...Consumer materialism means: (1) possessions as defining success; (2) the centrality of acquisition in one's life; and (3) acquisitions as the pursuit of happiness, consumer attaches to worldly possessions. Bearden identified three sub traits of materialism: (i) possessiveness; (ii) non-generosity; and (iii) envy and elaborates about materialism (the term often used in place of consumerism) as "belief in material growth as placing a high value on material comforts and conveniences, valuing economic effort, and viewing actions taken for the common good as possibly working against an individual, materialism as giving top priority to physical sustenance and safety.

Researching Individualism and Consumerism

The association between these two values is unique in the sense that both mutually depend upon each other. Capitalism by promoting individualism pressured existing family set-up and the purpose of doing so is market as discussed by researchers (Sarkar, 1997; Smith and Peterson, 1998). Smith and Peterson (1998: 264) commented, "television's audience was initially addressed as a family unit, but over time, influenced by advertising needs and changing social mores, this has changed so that increasingly groups and specialized interests are targeted more programmes (and channels) have emerged designed for sections of the audience" which in itself promote capitalist economy.

So, when individualism and consumerism both pressure each other there is a fundamental difference in which media consumption takes place on Western and non-western societies. "In Western cultures, media consumption is individualistic; in non-western cultures, it is responsible [and most often collective activity]" (Sitaram, 1991: 225).

Though various scholars especially in anthropology and social psychology have theoretically discussed individualism, little has been empirically tested. As discussed in Chapter 2, television-viewing impacts value orientation of its viewers. Many researchers have empirically studied consumerism nationally as well as internationally. More of international studies about consumerism have been conducted using case of People's Republic of China and Latin America.

International Research Studies about Consumerism

Many researchers have commented about association between television and consumerism. Wells (1972) comments, "television has fostered 'consumerism' in Latin America at the expense of 'producerism'. Television in America is linked to the US manufacturers through branches of major United States advertising agencies. As a result, most government in the region believe that commercial television is free whereas educational television is a luxury, thus preventing a developmental purposes (producerism)". Chandra (2000) notes "one of the gigantic swindles perpetuated on the public today is that commercial television is free (not so now), that advertising pays for the goods they purchase. Many television serials like Dallas, Dynasty and other prime-time soaps are themselves advertisements for wealth, luxury, fashion and high consumption lifestyles (defined as consumerism)". "Television technology is not neutral. It comes with a package, the heavy component of which is TV commercial" (Vilanilam, 1989: 495). "Television propagates a global consumer culture that is wasteful, unjust and environmentally unsound, and when this culture is put forward as the only one to aspire for it perpetuates economic disparities and unsustainable lifestyles" (Dixit, 1998).

Oliveira (1986) conducted micro research in Central American nation of Northern Belize, where television production is non-existent. Viewers watched television shows produced in the United States and Mexico. The research compared the relationship between exposure to US and Mexican television with the product preferences of people participating in the survey. Oliveira found that people who watched more US shows preferred more of US products and less of those produced in Central America. On the other hand people who watched more of Mexican shows preferred Central American products over those produced in United States. The research examined validity of satellite television and dependency theory and concluded that US programmes did impact the product preferences of television viewers.

Kottak (1990) in the context of his research in Brazil remarked that television is instrumental in providing information about products (their existence, appearance, function, availability and current prize) directly and regularly to people in small communities and remote areas. He noted, "new consumer patterns show up in hair styles, diet and clothing...A final example of TV impact on consumers' behaviour is a widespread correlation and probable negative public health effects. With its fashionable society women and powdered milk advertisements Brazilian television promotes early weaning and bottle-feeding" (1990: 162).

Allen (1992) examined the differential influences of news, telenovelas, and entertainment programmes on consumerism (or materialism) in media-rich, underdeveloped Venezuela through interviews with 305 samples in the age 15+, of different education and income levels, living in Arquisimeto/Cabudare. It is argued that adventure and crime stories and the telenovela situate action in a desirable materialistic framework, encouraging consumerism. Materialism is present in both domestic TV, through the telenovelas, and TV imported from the US, primarily adventure/crime stories. It was found that TV viewing promotes individual perceptions of the good life.

Sklair (1994) studied consumption and consumerism in China within a global system perspective and investigated the idea that a culture-ideology of consumerism is being created in urban China as it is in other parts of the non-capitalist and former communist worlds. From a sample of 591 viewers in Shanghai work units related to their attitudes about advertising and consumption, purchase of consumer durables, and knowledge of ten specific global brands. Groups were identified as pro-consumers, moderate consumers, and anti-consumers.

James (1995) conducted ethnographic research of cultural change in Hungary in the context of economic transformations of East-Central Europe. This opened the floodgates to sea of Western advertising and other symbolic activities aimed at informing the public about new products, their appropriate use, and their social and cultural meanings as markers of status. Ethnographic fieldwork was done in 1992-93 to explore how people in post-communist Hungary were making sense of this cultural transition. The research concluded that the significance of advertising and other marketing activities in such a setting lied not in their direct pitches to purchase this or that product, but rather in their subtle instruction in how to experience and express one's cultural identity through spending.

Zhao and Murdock (1996) explored the cross-cutting forces shaping the emergence of consumerism in contemporary People's Republic of China through a detailed research of the craze for the Transformer toys among Chinese children in 1989, drawing on interviews (sample size not specified) with children, observation of families and shops, fieldwork, and translations from Chinese sources. It is argued that rather than viewing the craze as evidence of cultural imperialism and Americanization, it should be understood as the product of a complex interplay between mass media and three cultural formations: the long-standing and still active legacy of Confucianism; past and present experiences of actually existing communism; and emerging potentials of the new market-oriented economy. Results suggest that cultural globalization is a highly uneven process that must be investigated with proper regard to the specific historical and social dynamics of particular situations.

Feldman (1998) reports about materialism and individualism as social attitudes of youth in Japan. It is reported that expansion of cable and satellite television has multiplied foreign news and information access. The research concludes that the loyalty to nation as abstract identity is declining and concern of younger generation has turned to focus mainly on the self.

Wei and Pan (1999) reported results from 1995 parallel cross-section surveys of probability samples from Beijing and Shanghai, addressing two questions: What consumerist value orientations are anchored in Chinese individualism? How is mass media exposure relate to the emergence of such values? Three stable consumerist value orientations were identified: conspicuous consumption, aspiration for self-actualization, and worshipping Western lifestyles. Those who were younger, better educated, and financially better off are the most likely to hold these values. Further, holding these values is related to frequent reading of consumer magazines and exposure to outdoor advertisements.

Indian Research Studies about Consumerism

Vilanilam (1989) observed that television advertising in India presents a heterogeneous cross-section of the Indian populace with images of products and lifestyles that were economically out of reach for the vast majority. In this way advertising constitutes an imaginary world of desire that addresses not only the elites that can purchase such goods, and in so doing affirm their elite status, but the masses of Indian workers and poor for whom the imaginaries is merely a reference point, a shaper of aspirations, an ideal model of success, pleasure, and envy.

Venkatesh (1995) discussed a new theoretical framework of ethno-consumerism-addressing consumerism as an ethno-cultural phenomenon as employed in comparative cultural studies of consumer behaviour. The research focused on technology adoptions and diffusion in India, such that behaviour could be studied both cross- and intra-culturally. Household technology was analysed within a cultural context, which reflected modernization and rising consumerism.

Grewal (1999) investigated role of Barbie products in the creation of new Indian consumers. He commented about the process of creating consumers for globalised products and the transformation of US culture into one characterized by consumption. He noted that although Indian sales of Barbie products during the 1980s were below Mattel's expectations, the increased number of diasporic Indians and production of Barbies wearing traditional Indian clothing might signify the emergence of a transnational market. Additional attempts by the Indian state to attract investments from transnational Indians and the creation of periodicals targeting non-resident Indians were additional factors.

Johnson (2000) in his research in two villages of Maharashtra found that television had played distinct role in the process of consumerism. "Villagers want to own many more things, which were once considered only luxuries, television itself being a prime example of this...there was a shift towards urban modelling, entry of T-shirts and wearing of baseball caps, change of hair-style amongst the girls, use of toothpaste and tooth brushes and even rinsing with mouthwash."

Though much of theoretical literature on values of individualism and consumerism has less empirically specificity, the association of consumerism and television has been empirically demonstrated through few studies conducted mainly in China and Latin America. Influence of television on value of individualism does not seem to have been empirically tested unlike consumerism though media text analysis indicated prevalence of both these values in Western programming.

Transnational Television Viewing

Present research defined transnational television as "viewing of English language television programmes or foreign origin channels". For the initial analysis both the levels of viewing were categorized separately but for further analysis both were consolidated and analyzed as transnational television 'Exposure' and 'No exposure'.

Graph 9.1 shows that little more than half (54.8 per cent) of the viewers were not exposed to transnational television at all wherein remaining had varied degree exposure. About 29 per cent of viewers either watched English language programmes or foreign origin channels. The viewers with exposure of transnational television were one-sixth (16.2 per cent) who watched transnational television in form of English language television programmes as well as foreign origin television channels.

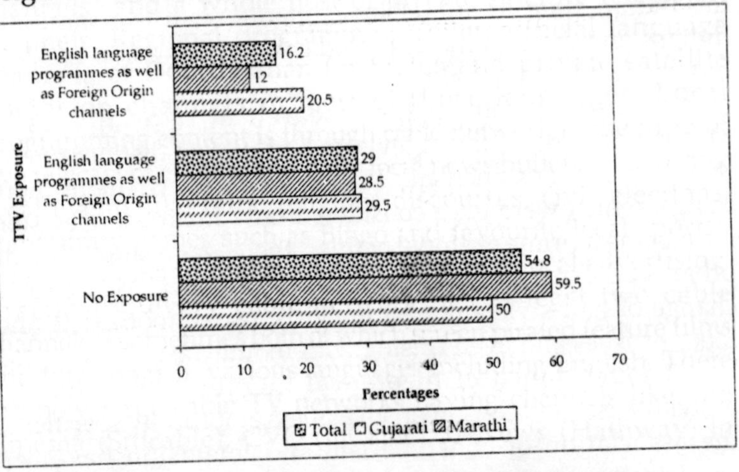

Graph 9.1: Nature of TTV Exposure by Mother tongue

Graph 9.1 clearly shows that Gujaratis exposure to transnational television was higher than that of Marathis. However, proportions of viewers watching transnational television in one of the forms were nearly thirty per cent (29.5 per cent Gujarati and 28.5 per cent Marathis) irrespective of their mother tongue. One-fifth (20.5 per cent) of the Gujaratis watched transnational television both in form of language as well as channel compared to only 12 per cent Marathis.

In order to get two clear categorizations of viewers with and without transnational television exposure, varied degree of exposure was clubbed together. Further analysis was carried out with independent variables like sex, age, education, occupation and SES. Fewer females were watching transnational television compared to males. Exactly half (50 per cent) of the males watched transnational television as against 40.5 per cent of females. More than half (57 per cent) of Gujarati males were watching transnational television compared to exactly same proportion of Marathi males who were not watching transnational television.

Table 9.2 shows that seven out of ten Gujarati viewers in 20 to 30 years of age group were watching transnational television compared to 42 per cent of their Marathi counterparts. Proportion of Marathis viewers in 30 to 40 years age group watching transnational television (44.2 per cent) was somewhat comparable to 20 to 30 years of age group (40.5 per cent) unlike the difference among Gujarati viewers. Lesser proportion of Gujarati viewers in 30 to 40 years age group were watching transnational television (42.9 per cent) compared to 70.3 per cent of those in 20 to 30 years of age. It seems that age and transnational television viewing had an inverse relationship.

Table 9.2: Percentage Distribution of Transnational Television Viewing and Age by Mother tongue

TTV	*Gujarati*				*Marathi*				*Total*		
Viewing	*20-30*	*30-40*	*40-50*	*Total*	*20-30*	*30-40*	*40-50*	*Total*	*20-30*	*30-40*	*40-50*
No Exposure	29.7	57.1	77.4	50.0	58.0	55.8	66.7	59.5	45.3	56.6	72.4
Expo sure	70.3	42.9	22.6	50.0	42.0	44.2	33.3	40.5	54.7	43.4	27.6
Total	100.0	100.0	100.0	100.0	100.0	100.0	100.0	100.0	100.0	100.0	100.0

Table 9.3 suggests that level of education of viewers had direct association with transnational television viewing regardless of mother tongue. Almost 67.2 per cent of post-graduates watched transnational television compared to only 27.6 per cent of school educated (SSC/HSC and below) viewers. Only about one-third (32.8 per cent) of the post-graduates were not exposed to transnational television compared to about 46.2 per cent of the undergraduates.

Table 9.3: Percentage Distribution of Transnational Television Viewing and Education by Mother tongue

Mother tongue	*Gujarati*			*Marathi*			*Total*		
Education TTV Viewing	*SSC/ HSC and below*	*UG and Graduates*	*Post Graduates*	*SSC/ HSC and below*	*UG and Graduates*	*Post Graduates*	*SSC/ HSC and below*	*UG and Graduates*	*Post Graduates*
No Exposure	79.5	45.3	30.8	82.3	47.3	35.7	81.3	46.2	32.8
Exposure	20.5	54.7	69.2	17.7	52.7	64.3	18.7	53.8	67.2
Total	100.0	100.0	100.0	100.0	100.0	100.0	100.0	100.0	100.0

Tale 9.4 shows that occupation of the viewers also influenced transnational television viewing with differences across mother tongue. Close to seven out of ten students were watching transnational television irrespective of their mother tongue. Among Gujaratis little more than half (53.1 per cent) of the viewers in service/professionals and 45.3 per cent of viewers in business were watching transnational television unlike Marathis. More than one-third of housewives across mother tongue were also watching transnational television.

Like education of the viewers, SES of the viewers also had direct association with transnational television viewing irrespective of their mother tongue. Higher the SES, higher was the transnational television viewing. Six out of ten SES-A viewers irrespective of their mother tongue were viewing transnational television compared to about one fifth of SES-C viewers. Interestingly more than half (56.4 per cent) of the SES-B Marathis viewers were watching transnational television compared to 42.5 per cent of Gujarati SES-B viewers indicating that higher proportion of Marathi middle SES compared to Gujarati middle SES were viewing transnational television.

Table 9.4: Percentage Distribution of Transnational Television Viewing and Occupation by Mother tongue

TTV Viewing	*Gujarati*				*Marathi*			
	House wife	*Service/ Professional*	*Business*	*Student*	*House wife*	*Service/ Professional*	*Business*	*Student*
No Exposure	61.8	46.9	54.7	25.0	63.2	60.5	78.3	32.0
Exposure	38.2	53.1	45.3	75.0	36.8	39.5	21.7	68.0
Total	100.0	100.0	100.0	100.0	100.0	100.0	100.0	100.0

Independent sample t-test scores of transnational television viewing and mother tongue showed that there was a statistically significant difference in transnational television viewing among Gujaratis and Marathis. These differences can be explained due to differences in terms of education, SES and occupational pattern of viewers across mother tongue as discussed above.

Scores of Individualism and Consumerism

As elaborated in Annexure D, cumulative scores were calculated from the scales on individualism and consumerism. Individualism scale comprising of statements on individualism and familialism while consumerism scale had statements related to consumerism and statements not related to consumerism [*See* Annexure A(i) and A(ii)]. Table 9.4 shows that mean score and variance analysis of individualism had marginal difference but the variance in the value scores of Gujaratis was higher than Marathis. This indicates that the degree of individualism (in form of scores) across mother tongue did not vary but within Gujarati viewers the variation in individualism scores was higher compared to Marathis. Similarly Table 9.5 also shows the statistics of consumerism scores among Gujaratis and Marathis. Interestingly unlike individualism the scores of consumerism had no difference across mother tongue even in terms of variance. The mean values also did not differ across mother tongue suggesting that the degree of consumerism among Gujaratis and Marathis did not differ.

Table 9.5: Scores of Individualism and Consumerism by Mother tongue

Statistics	*Individualism*		*Consumerism*	
	Gujarati	*Marathi*	*Gujarati*	*Marathi*
Mean	59.68	59.96	62.82	62.83
Std. Error	0.56	0.49	0.86	0.89
Median	60.00	61.00	63.00	63.00
S.D.	7.98	7.00	12.19	12.59
Variance	63.63	48.95	148.51	158.54
Minimum	38.0	40.0	35.0	36.0
Maximum	83.0	80.0	105.0	96.0
Range	45.0	40.0	70.0	59.0

In order to examine statistical significance of difference in the scores, independent sample t-test was employed for individualism as well as consumerism. There was no significant difference between the values of individualism as well as consumerism among Gujaratis and Marathis. This indicates similarity of value system within Indian cultural context. In spite of differences in mother tongue, education, SES and occupation viewers had no significant difference in values.

Correlation of scores of individualism and consumerism was calculated so as to ascertain association between two sets of values. Values of individualism and consumerism were found to be significantly correlated (r=0.422, p=0.05) suggesting that the viewers having higher degree of individualism tend to have higher degree of consumerism and *vice versa*. The finding supported the review of literature that both the values are mutually dependent on each other (Sarkar, 1997; Smith and Peterson, 1998).

Transnational Television, Individualism and Consumerism

In spite of the fact that transnational television viewing varied across mother tongue, there was no statistically significant difference in the value scores across mother tongue. Table 9.6 indicated that if data are pooled together 72.3 per cent of transnational television viewers had 'medium' to 'high' individualism scores compared to 62.5 per cent of viewers who were not watching transnational television at all. Proportion of 'low' individualism scores was 37.4 per cent among 'no exposure' compared to 27.6 per cent of transnational television viewers. In brief, it seems that transnational television viewers tend to have higher degree of individualism.

Table 9.6: Percentage Distribution of TTV Viewing by Individualism Scores

Individualism	*Transnational Television Viewing*		*Total*
	No Exposure	*Exposure*	
Low	37.4	27.6	33.0
Medium	35.6	39.2	37.3
High	26.9	33.1	29.8
Total	100.0	100.0	100.0

Table 9.7 shows transnational television exposure and consumerism had some kind of relationship. If data are pooled together, 71.2 per cent of transnational television viewers had medium or high degree of consumerism score compared to

Table 9.7: Percentage Distribution of TTV Viewing by Consumerism Scores

Consumerism	*Transnational Television Viewing*		*Total*
	No Exposure	*Exposure*	
Low	34.7	28.7	32.0
Medium	39.3	33.1	36.5
High	26.0	38.1	31.5
Total	100.0	100.0	100.0

65.3 per cent of viewers not watching transnational television at all. Similar to individualism, proportion of transnational television viewers with 'low' consumerism was 34.7 per cent compared to 28.7 per cent of non-viewers.

It was felt necessary to analyse the background of viewers of transnational television keeping in mind the earlier discussion. Hence, transnational television viewing was examined in relation to non-television media usage and English language background of the viewers.

English Language Background, Transnational Television and Values

Since transnational television viewing predominantly referred to English language, viewer's usage of other non-television English language media in form of newspaper and magazine, films, plays and Internet were examined. Close to nine out of ten (89 per cent) of the viewers who watched transnational television were using at least one other English language media. At the same time more than one-third (37 per cent) of viewers who used non-television English media were not watching transnational television and only 11 per cent of viewers who were not using English language media but were watching transnational television. Almost four out of ten (39.5 per cent) of the total viewers were not using non-television media in English while as discussed earlier more than half (54.8 per cent) of the viewers were not watching transnational television.

Little less than half (48 per cent) of Gujaratis and more than three-fourth (75.6 per cent) of Marathis were not at all using English language media. About half (52 per cent) of Gujaratis and one-fourth (24.4 per cent) of Marathis were using at least one English media other than TV. It was found that transnational television viewers most frequently accessed other non-television English language media.

In order to examine the role of English language background of viewers on transnational television viewing, responses of three questions: "language of conversation at home", "medium of instruction at school" and "usage of

English language non-television media" were consolidated. After initial analysis the data were pooled together and two categories were created, namely 'English language background' and 'non-English language background' of the viewers.

Table 9.8 shows that less than half (47.5 per cent) of Marathis and one-fourth (26 per cent) of Gujaratis had no exposure to English at school, at home or in other non-television media. About one-tenth (8.5 per cent) of Gujaratis were exposed to English at all three levels compared to its half (4 per cent) that of Marathis. Interestingly about one-third of viewers were exposed to English at either of the levels irrespective of their mother tongue.

Table 9.8: Percentage Distribution of English Language Background of Viewers and Transnational Television Viewing

English language background	*Mother tongue*		*Total*
	Gujarati	*Marathi*	
Not Applicable	26.0	47.5	36.8
Either level*	32.0	35.5	33.8
Either of the two levels	33.5	13.0	23.3
All Three levels	8.5	4.0	6.3
Total	100.0	100.0	100.0

* English as a medium of instruction at school, language of conversation at home, usage of non-television media in English language.

Independent sample t-test was employed to examine statistical significance of difference in English language background and mother tongue which showed that the English language background of the viewers was significantly different across mother tongue.

In the light of earlier discussions, it is clear that the values of individualism and consumerism did not differ but transnational television viewing as well as English language background across mother tongue differed statistically significantly. It was also observed that with English language background there is an increase in individualism score indicating direct relationship of English language background with value of individualism. Similarly, with English language background there is an increase in consumerism score indicating direct relationship of English language background with value of consumerism.

Since English language background influenced values of the viewers, its association with transnational television viewing was also examined. Nine out of ten viewers who watched transnational television had English language background. This indicates that transnational television viewing was influenced by the socio-cultural factors as well as other media habits and prior exposure to English language. In order to examine interplay of various independent and dependent variables correlation analysis was carried out.

Association of Variables

Bi-variate correlation of the value scores was examined with transnational television viewing. Table 9.9 clearly shows that transnational television viewing was significantly correlated with values of individualism and consumerism suggesting that higher the transnational television viewing higher is the scores of individualism and consumerism.

Correlation of television related variables such as amount of time spent on television and years of cable connection with values of individualism and consumerism was suggested that amount of time spent on television was directly and significantly correlated with individualism and consumerism scores.

Table 9.9 also shows that years of cable connection was directly associated with transnational television viewing (r=0.172) and inversely (though not significantly) correlated with scores of consumerism (r=–0.0105). Amount of time spent on television daily was significantly correlated with the values of individualism (r=0.137) and consumerism (r=0.176). Similarly, years of cable connection and daily time spent on television were inversely correlated though not statistically significant suggesting that with more number of years of cable connection, the amount of time spent on television reduced.

Table 9.9: Correlation of Television Related Variables and Value Scores

		Individualism	*Consumerism*	*TTV Viewing*	*Daily TV time*	*Years of Cable connection*
Individualism	r	1.000	.422**	.172**	.137**	.008
	Sig.		.000	.001	.006	.879
Consumerism	r		1.000	.160**	.176**	-.105**
	Sig.			.001	.000	.036
TTV Viewing	r			1.000	-.014	.263**
	Sig.				.776	.000
Daily TV time	r				1.000	-.061
	Sig.					.225
Years of Cable connection	r					1.000
	Sig.					.

r = Pearson Correlation . ** Correlation is significant at the 0.05 level (2-tailed).

In order to explore association of transnational television viewing, other television related variables and value scores of individualism and consumerism with independent variables like age, sex, education, occupation and SES have been presented in form of correlation matrix in Table 9.10 which reconfirms the earlier conclusion by indicating significant and direct correlation between transnational television viewing and sex (r=0.120), education (r=0.373) and SES (r=0.414) and inverse correlation with age (r=–0.217). Sex was also significantly correlated with Individualism (r=0.131) as well as consumerism (r=0.229) scores while age was inversely correlated with consumerism (r=–0.329).

Table 9.10: Correlation of Value Scores, TTV Viewing and Independent Variables

		Individualism	*Consumerism*	*TTV Viewing*	*Age*	*Sex*	*Education*	*Occupation*	*SES*
Individualism	r	1.000							
	Sig.	.							
Consumerism	r	.422**	1.000						
	Sig.	.000	.						
TTV Viewing	r	.172**	.160**	1.000					
	Sig.	.001	.001	.					
Age	r	-.078	-.329**	-.217**	1.000				
	Sig.	.121	.000	.000	.				
Sex	r	.131**	.229**	.120**	-.045	1.000			
	Sig.	.009	.000	.016	.373	.			
Education	r	.064	.026	.373**	.033	-.029	1.000		
	Sig.	.199	.606	.000	.510	.560	.		
Occupation	r	-.061	-.094	.077	-.145**	-.403**	-.117**	1.000	
	Sig.	.225	.061	.125	.004	.000	.020	.	
SES	r	.048	-.015	.414**	.132**	-.146**	.587**	.098	1.000
	Sig.	.340	.767	.000	.008	.003	.000	.051	.

r = Pearson Correlation. ** Correlation is significant at the 0.05 level (2-tailed).

Table 9.11 shows the correlation of English language exposure and value scores and as evident from earlier discussion, transnational television viewing was significantly correlated with other English media usage and English language background. English media usage as well as English language background had significant correlation with value scores of individualism and consumerism.

It may be concluded that irrespective of mother tongue, scores of individualism and consumerism were not only influenced by transnational television but also by other media behaviour and socio-cultural background of viewers. Demographic and social variables like age, sex, education, SES, English media usage and English language background influenced transnational television viewing. Even years of cable connection influenced amount of time spent on television and consumerism scores inversely. This may mean that as the years of cable connection increases; viewers tend to watch less amount of television. Or it may also mean, "negative attitudes towards television increase with higher income and with years of exposure...[as] length of exposure increases, clear statistical correlation between television-viewing and other variables diminishes" as found by Kottak through his study in Brazil (Kottak, 1990: 140).

Table 9.11: Correlation of Value Scores and English Media Viewing

		Mother Tongue	*English language back-ground*	*Trans-national TV Ex-posure*	*Non-TV English Media*	*Indivi-dualism*	*Consu-merism*
Mother Tongue	r	1.000	-.223**	-.095	-.236**	.018	.001
	Sig.	.	.000	.057	.000	.714	.990
English language background	r		1.000	.495**	.953**	.166**	.138**
	Sig.		.	.000	.000	.001	.006
Trans-national TV Exposure	r			1.000	.511**	.138**	.135**
	Sig.			.	.000	.006	.007
Non-TV English Media	r				1.000	.153**	.125**
	Sig.				.	.002	.013
Indivi-dualism	r					1.000	.422**
	Sig.					.	.000
Consu-merism	r						1.000
	Sig.						.

r = Pearson Correlation. ** Correlation is significant at the 0.05 level (2-tailed).

Therefore, though the values of individualism and consumerism have been associated with transnational television viewing, the viewing itself depends upon other variables. Mother tongue *per se* does not influence values of individualism and consumerism but it does indicate differences in transnational television viewing due to difference in other variables.

During Family Group Interviews researcher came across comments about English language television viewing in three Gujarati and one Marathi households (Patels, Ashars, Kadakiyas, and Baligas), Patels father (65 years) and Ashars Son (10 years) had a similar comment about English language television viewing. Mr. Patel felt that his sons watch a lot of English language television, which had made them aggressive. They now argue a lot and do not accept others point of views. Patel's Parents were mainly into Gujarati serials and spiritual channels and they expressed that whatever shown on television is correct and it teaches us lessons for life. But both were unhappy about their sons English Television programme viewing. The Mother also expressed that she did not understand content watched by the children on the television. The parents get to watch whatever they like only when their sons are away.

Ashars 10 years old boy who was a regular viewer of English language television, when asked how he thinks television influences him he said, "I have a feeling I have become aggressive". He also stated that in his school, children talk a lot about television shows and his parents do not like him to watch television because *'they had never watched it when they were young and so they feel jealous of me'*. When asked if there was anything besides television that the child enjoys the most, cycling was his choice. Mr. Ashar felt that his son's English had improved because of television. The mother reported that the child remains alone many times at home and watch television; he asked her one day why divorces happen and what is marriage. The child mentioned, *"what I watch on television and what my life is both are very different. I do not know what is good or bad but it is fun to watch it"*.

Mrs. Baliga at Baligas household was concerned about her younger daughter of 12 years who watch television alone many times. She shared an incident that her younger daughter after returning from a stay at their relative's house started watching 'Friends' (English daily soap). Mother allowed her to watch it assuming it was about friendship, which was stopped recently. Mrs. Baliga felt that what her daughter learns from programmes like *Friends* 'is not good' but she was not able to specify what was not good.

Kadakiyas were actively involved in Chinmaya Mission activities; two unmarried daughters were working in banks, younger one in multinational bank and elder daughter in nationalized bank. The younger one was mainly into English films on television. She stated, "viewing of English films on television shows very different culture and can make people self-centred. There is lot of sex and violence also. I watch it to see the way they tell stories and their techniques". She felt that if one views without using one's mind, it could influence viewers.

The grandson at Velunkars expressed that the influences of television is visible in the college campus. The students discuss a lot about television programmes. His mother commented that his studies got influenced because of his TV viewing. He scored 87 per cent in 12th standard and had gone down to 62 per cent in Engineering. Even his clothing style, dancing, etc. had changed and he was a part of a dance-music group. He was a heavy consumer of English programming on television which he watched alone as discussed earlier.

In Brief

Present research revealed that transnational television exposure was limited to about two-fifth of viewers indicating higher preference for Indian/Indian language television programmes compared to transnational television. In spite of differences in material conditions of Gujarati and Marathi households, there were similarities in value orientations of individuals. At the same time, within Indian cultural setting mainly due to socio-economic differences, the patterns of media consumption as well as transnational television viewing varied significantly.

Education and SES were found to be associated with viewing of transnational television irrespective of mother tongue of the viewers. Gujaratis in 20 to 30 years of age group and Marathi viewers irrespective of their age were watching transnational television. Students and Gujarati businessmen watched more of transnational television.

Independent sample t-test suggested that transnational television viewing varied across mother tongue. At the same time in spite of education, occupation and SES differences across mother tongue, there was no significant difference between values of individualism and consumerism among Gujaratis and Marathis. Viewers belonging to English language background were found to be having higher scores of individualism and consumerism.

Family group interviews revealed that viewers of transnational television irrespective of their age understand and articulate that there was a difference in the cultural values portrayed on television and their own lives but they found it interesting to view such programmes. Viewers in the age range from 10 to 65 years expressed fear that transnational television was making viewers 'aggressive' and may be influencing traditional Indian family hierarchy.

Value scores of individualism and consumerism were found to be significantly correlated with each other as well as transnational television viewing. Correlation matrix of independent variables with individualism and consumerism revealed that age was inversely correlated with consumerism. Though transnational television viewing influenced the values of individualism and consumerism, transnational television viewing itself depended upon other variables like age, sex, education, SES and even mother tongue.

10

Television, Culture and Individual

Based on the previous analysis and discussions, it can be concluded that the cry for cultural crisis and erosion of values in Asian cultures in general and Indian society in particular is a mere 'cry' without any clear evidences. Transnational television viewers are members of the society who already have got exposed to Western value system due to their education, socio-economic status and household environments and, therefore, transnational television has little or no role to play in transforming cultural milieu of their societies.

Indian colonial past has made English second language for many Indians but unlike print media usage in English, audio-visual media consumption still takes place in Indian languages as found through the present research.

Present research findings do not support the premise of cultural-media imperialism. No evidence was found to support that the transnational television has swept the local culture or language preferences of viewers. The concept of media imperialism did not anticipate rapid growth and expansion of national and regional language television as witnessed in India and elsewhere. Indian language television channels or many channels of Indian-origin continue to be most popular as compared to transnational channels. Transnational television channels were forced to 'localise' their programming if not language in few cases as witnessed in India and elsewhere in the world.

In spite of fears of media imperialism actual consumption of transnational television continue to be limited (less than ten per cent). Viewers in the 20 to 30 years of age group, mostly students, having English language background were found to be consuming transnational television more than other category of viewers. At the same time programme preferences appear to be changing with age, which means that those watching transnational television may change their preferences in future. At the same time younger generation of Gujaratis do not use media in their mother tongue add complexity to the situation.

The research also confirms conceptual premises of "active audience" and "polysemy" and provides evidences for "resistance". Family group interviews and individual interviews suggested that influences of transnational television get acknowledged by the viewers. Media text (in form of television programme) was interpreted by the viewers depending upon their demographic and socio-cultural characteristics.

Half of the viewers by responding that they had not viewed any transnational television programmes unsuitable to themselves and their family indicate 'silence'. Even the influences in private and public spheres were found to be accepted mainly with reference to personal self and not on family or society suggesting negation of influence by the viewers.

Individual television viewer is located within the collective framework of Indian family. Though multiple television sets are being ordered, individual's cultural values are located within 'familial collectivism' and their education, occupation, other media usage as well as their age and sex has influence on their usage of transnational television. While the collective mother tongue cultural traits influence primary and secondary variables of individual viewers, transnational television is not the only variable to influence the worldview of the individual. For the large majority of the individuals television is not yet transnational.

It would be difficult to conclude from the present research that transnational television promote values of individualism and consumerism because today, "transnational media coexist with domestic, and compete with viewers; domestic production can become even more commercial, garish and explicit than the western originals" (Sreberny-Mohammadi, 1991: 181).

In such circumstance, cultural-media imperialism, if any, is taking place in a distinct way. The evidences suggest that viewers having English language background (with English education, language at home and usage of non-television English media) were viewing transnational television more frequently and had higher degree of individualism and consumerism. But large part of television viewing population is untouched in terms of exposure to transnational television directly. The indirect pressures in form of presence of multiple television sets, conflicting viewing priorities and diverse viewing preferences, negotiations for setting time table, questioning of the patriarchal hierarchy are some of the indicators of change in Gujarati and Marathi households, and so Indian culture due to transnational television.

ANNEXURE-A (i)

List of Statements Individualism—Familism/Collectivism

No.	*Statements 'For' Individualism*
1.	Individual is the most important element in any society.
2.	I do not believe in interfering with other people's affairs.
3.	I am successful due to my own efforts.
4.	If I am earning money I have a right to decide about how to spend them.
5.	The couple has a right to decide when to have children.
6.	My marriage partner has to be decided by me not by my family.
7.	Sons should separate from the parents after marriage.
8.	My needs are more important than others' needs.
9.	I make it a point to talk about my capabilities so that others know about it.
10.	I hate to share things with my brothers/sisters.
11.	I value being alone than to go for a social visit.
12.	If both the partners are not happy they should take divorce.
13.	One should live one's life independent of the society.
14.	Women can have equal rights as men.
15.	Woman's future does not get decided after marriage.
	Statements about Familism/Collectivism 'Against' individualism
16.	Community interests are above the individual interests.
17.	My family support is very important to me.
18.	My success can be due to blessings of my family.
19.	I hate to disagree with the group members.
20.	I would sacrifice activity I enjoy most if my family does not approve of it.

ANNEXURE-A(ii)

List of Statements: Consumerism

No.	*Statements related to Consumerism*
1.	I make it a point to keep up with the latest fashion labels.
2.	I believe physical comforts always give happiness.
3.	I think family's status is decided by what they own rather than by their family values.
4.	Shopping gives me lot of pleasure.
5.	Many things bought in my family are for status symbol and not for use.
6.	I prefer to go for the vacation at the holiday resort than to visit my relatives at my native place.
7.	I always buy branded things even if they are expensive.
8.	Today goods you use do not have mere use value but are personality statements.
9.	After seeing so many advertisements on TV, I get tempted to buy goods.
10.	There is nothing wrong in speculative earnings ultimately what counts is money.
11.	Life is for given once, I believe in enjoying it to the fullest.
12.	It is fine to spend on cosmetic surgery to look beautiful/handsome.
13.	Money can buy you everything including happiness.
14.	I am not bothered about tomorrow, today is most important day of my life.
15.	A successful person is one who earns lot of money.
16.	If some product is very expensive I do not mind buying it at instalment and start enjoying the benefits today.
17.	Fast food gives me more satisfaction than traditional meals.
18.	Life is worthless if we do not have goods and lifestyles shown in TV.
19.	I believe exchange offers are the best bargains to get rid of old things.
	Statements Not related to Consumerism
20.	The present approach to get quick money in life is contrary to our values.
21.	'Prizes' and 'buy one get one free' offers unnecessarily tempt you to buy more.
22.	I look for value for money whenever I buy new things.
23.	Wealth is god given for the service of others.
24.	Things that make me happy usually come without money.
25.	I always carry water from home than to buy new bottle every time.

Operational Definitions

- **Transnational Television:** Transnational television has been defined as foreign-origin television channels (whose ownership as well as content of programming is not Indian) and have television programmes in English language.
- **Household:** A household is a group of persons who commonly live together and would take their meals from a common kitchen unless the exigencies of work prevent them from doing so. The member of the household may be related by blood, unrelated or having a mix of both.
- **Family:** Family for the present research means persons residing in the household who are related to each other by blood or marriage. It has been used interchangeably for household.
- **Gujarati Speaking Household:** Household members having Gujarati language as their mother tongue and converse primarily in Gujarati.
- **Marathi Speaking Household:** Household members having Marathi language as their mother tongue and converse primarily in Marathi.
- **Viewer:** Any individual of either sex residing in Cable and Satellite television household in the City of Mumbai between the age group of 20 to 50 years. The term also has been used interchangeably with viewer.
- **Individualism:** The person's strong sense of individual identity as opposed to family, self-assertion and a belief that he/she is primarily responsible for his/her life as against social obligations and interdependence to the family and community.
- **Consumerism:** Consumerism is an urge to acquire and possess material goods, money as criterion for success and high value attached to physical pursuits.

- **Life-style Products:** Possession of products like Mobile phone, credit card, laptop computer and club membership have been defined as life-style products.
- **English Language Background:** Refers to those who have studied in English medium school, and/or converse in English at home and/or use non-television media in English language.
- **Non-Television Media:** Refers to media other than television (mainly newspaper, magazines, books, radio, films and Internet).

Construction of Indices

Reason for construction of indices was better presentation and meaningful interpretation of the findings. Following indices were used for analysis and interpretation.

- Socio-Economic Status (SES)
- Non-Television Media Time
- Daily Average Television Viewing Time
- Language Efficiency
- Individualism
- Consumerism

Socio-Economic Status: Weighted mean was calculated for ten household appliances (telephone, two-in-one, music system, VCR/VCP/VCD, washing machine, fridge, two-wheeler, car, air conditioner, computer) and four life-style products (Lap-top, credit-card, club membership, mobile phone). In addition, points were given for multiple television set ownership, type of television set, education of head of the family and monthly family income. For weighted mean formula $[(N-n_i))/N]$ was applied where N= 400 and ni = Number of viewers having ownership of respective appliance. In all 18 items were considered to obtain Socio-Economic Status scores in three categories as classified in Table I.

Table I: Socio-Economic Status Indices Classification

Score	*SES*
4.18 to 10.03	SES-C
10.04 to 14.48	SES-B
More than 14.49	SES-A

The cumulative scores so obtained ranged from 4.18 to 28.19. The mean value of SES was 12.69 and SD was 4.81 (rounded till two decimals).

Media Time Spent: The viewers were asked about the amount of time they spent on four media: Newspaper, Magazine, Book, and Radio. Frequency of use of a particular medium was coded as shown in Table II. Other media like computer, Internet and VCR/VCP/VCD were not considered as for the majority it did not constitute 'daily media usage'.

Table II: Coding of Frequency of Media Usage

Frequency of use	*Code*
Everyday	7
Five-Six days a week	6
Three-four days a week	4
Once or twice a week	2
Once or twice a month	1

Formula for "Average Daily Media Time Spent" on non-television media was: [(Average minutes spent * Frequency of use of that media)/60]/7. Cumulative score for four media (newspaper, magazine, books and radio) was calculated so as to compare it along with television time usage.

Daily Average Television Viewing Time: Since the question of amount of time spent on television is a difficult one to get 'correct answer' the viewers were asked to report weekday and weekend time spent in morning-afternoon-evening and night watching television and amount of time spent each time. Daily average viewing time was calculated using following formula:

Daily Average Viewing Time =

[(Weekday hours * 5)+ (Weekend hours * 2)]/7

Some viewers had reported that they did not watch television on Sunday; in such cases weekend hours were multiplied with one instead of two. Minutes were divided by 60 to get hours. The range turned out to be half an hour to 11 hours a day. Index was constructed for further analysis has been shown in Table III.

Table III: Daily Time Spent on TV Indice Classification

No. of Hours Spent Per Day	*TV Time Spent*
Up to 2.50 hours	Low
2.51 to 4.00 hours	Moderate
More than 4.00 hours	High

Language Efficiency: The viewers were asked to report their familiarity with languages and degree of efficiency in terms of reading, writing, and speaking for each language. Except 0.05 per cent viewers, all were bilingual or trilingual. The 400 viewers reported varied language efficiencies. In total, familiarity with 11 languages (Gujarati, Marathi, Hindi, English, French, Russian, German, Bengali, Tamil, Konkani and Kannada) and one dialect (*Kutchi*) was reported.

Language familiarity data were coded into two ways. 'Familiarity with English' and 'Language Efficiency'. Familiarity with English was examined since transnational television was also defined in the context of English language television programmes. Viewers having reported ability to read, write or speak English were given score of 1 and those having English language familiarity at any level were scored zero. As much as 90 per cent of Gujaratis and 74 per cent of Marathis reported 'familiarity with English'.

Language efficiency scores were calculated by assigning scores to Reading (2), Writing (3) and Speaking (5) for each language. Person who was able to read, write and speak a particular language had total score of 10 for a particular language. Speaking was given higher weightage because television being audio-visual medium has little to do with ability to read or write.

Language efficiency scores so calculated ranged from 5 to 45 indicating familiarity with as many as five languages by few of the viewers. The mean score was 29.87 and SD was 8.72. Four Marathi viewers were illiterate and so were unable to read or write but they got higher scores due to their ability to speak more than one language. The language efficiency indice so constructed was classified in three categories as shown in Table IV.

Table IV: Language Efficiency Indice Classification

Scores	*Language Efficiency Level*
0 to 20	Low
21 to 35	Medium
36 and Above	High

Degree of Individualism: Responses to the 'for' statements on individualism were assigned scores as shown in Table V. The statements reflecting familialism were coded as 'against'. Annexure A(i) lists down both the types of statements from the individualism scales. Cumulative scores were converted into 'degree of individualism' as shown in Table VI. The individualism scores ranged from 38 to 83 and the mean score was 59.82 and SD was 7.49.

Table V: Scoring pattern for Value Scales on Individualism and Consumerism

Response	*Scores*	
	'For'	*'Against'*
Strongly Agree	5	1
Agree	4	2
Neutral	3	3
Disagree	2	4
Strongly Disagree	1	5

Table VI: Individualism Indice Classification

Score	*Degree of Individualism*
Up to 56	Low
57 to 63	Medium
Above 63	High

Degree of Consumerism: Responses to the 'for' statements on 'consumerist attitude' were assigned scores as shown in Table V. The statements reflecting non-consumerist attitude were coded as 'against'. Annexure A(ii) lists down both the types of statements from the consumerism scales.

Cumulative scores were converted into 'degree of consumerism' as shown in Table VII.

Table VII: Consumerism Indice Classification

Score	*Degree of Consumerism*
Up to 56	Low
57 to 68	Medium
Above 68	High

The consumerism scores ranged was from 35 to 105 and the mean score was 62.82 and SD was 12.38.

ANNEXURE-D(i)

Names of Media Brands Accessed by Viewers
List of Newspapers

1. Afternoon
2. Asian Age
3. Business Standard
4. Economic Times
5. Financial Times
6. Gujarat Samachar
7. Indian Express
8. Jame-e-jamshed
9. Janmabhumi
10. Lokmat
11. Loksatta
12. Mahanagar
13. Maharashtra Times
14. Mid Day
15. Mid Day (Gujarati)
16. Mumbai Chufer
17. Mumbai Samachar
18. Mumbai Sandhya
19. Navabharat Times
20. Navakal
21. Navshakti
22. Ratnagiri Times
23. Sakal
24. Samkalin
25. Sarnna
26. Sandesh
27. Sandhyakal
28. Sandhyanand
29. The Times of India + Bombay Times
30. Vartahar
31. Yshokarmi

ANNEXURE-D(ii)

List of Magazines

General Interest

1. Abhiyan
2. Asia Week
3. Chitralekha
4. India Today
5. Lok Prabha
6. Newsweek
7. Outlook
8. Prapanch
9. Readers' Digest
10. Saptahik Sakal
11. Shree Ank
12. Time
13. Week

Literary

1. Aawaz
2. Akhand Anand
3. Dakshata
4. Jatra
5. Kathashree
6. Marmik
7. Naval Prakash
8. Navaneet Samarpan

Business

1. A and M
2. Business India
3. Business Today
4. Business World
5. CA Institute Journal
6. Flight International
7. Forbes
8. Fortune
9. Intelligent Investor

Film

1. Cine Blitz
2. Film Fare
3. Jee
4. Stardust

Women's Magazine

1. Charchudhi
2. Chatura
3. Cosmopolitan
4. Elle
5. Femina
6. Grih Shoba/Shobhika
7. Gruh Lakshmi
8. Madhuri
9. Maher
10. Sahiyar
11. Sakhi
12. Society
13. Stree
14. Women's Era

Religious

1. Baudh Ani Vyachadharmi
2. Chinmayanand Mission
3. Gajanand Ashish
4. Jankalyan
5. Natvagayan
6. Saileeleela Mrut
7. Sangha Sahitya
8. Yugshakti

Special Interest

1. Amrut
2. Apparels
3. Auto India
4. Better Photography
5. Chandan Chakram
6. CHIP
7. Cloth Line
8. Computer Express
9. Computer Today
10. Debonair
11. Digit
12. Economist
13. Express Textiles
14. Health Positive
15. Inside Outside
16. JLT
17. Life Positive
18. MAD
19. Marwar
20. Network Solution
21. Nutrition and Health
22. Safari
23. Shamrdh Aai
24. Shatkar
25. Spardha Pariksha
26. Sports Line
27. Sports Star
28. Traveller
29. Tourism Express

ANNEXURE-D(iii)

List of Television Channels

Type of Channels	*Names of the Channels*
	General Entertainment
Indian	Zee, Sahara, SAB, B4U, Doordarshan
Indian origin Foreign ownership	STAR, Sony
Foreign	Zee English, STAR World, Sony Max
	Music
Indian	CMM, CCC, ATN, ETC, B4U
Foreign	MTV, Channel V
	Film
Indian	Zee Cinema, CVO, B4U
Foreign	STAR Movies, STAR Gold, HBO, AXN
	Sports
Indian	DD Sports
Foreign	STAR Sports, Ten Sports, ESPN
	News
Indian	DD News, Zee News, Aajtak
Indian origin Foreign ownership	STAR News, CNBC
Foreign	BBC World, CNN
	Others
Regional (Indian)	Alpha Marathi, Sahyadri, Taravahini, Etv, Prabhat, Alpha Gujrati, Gurjari, DD Bangla
Religious (Indian)	Astha, Sanskar, Sanskriti
Info-tainment (Foreign)	Discovery, National Geographic, Animal Planet
Others (Foreign)	Cartoon Network
Local cable network	in-mumbai, my channel, itv, cable film channel

ANNEXURE-E

Selected Tables

Table 1: Percentage Distribution of Language Preferences of Print Media by Mother tongue

Print Media Language	*Newspapers*		*Magazines*		*Books*	
	Gujarati	*Marathi*	*Gujarati*	*Marathi*	*Gujarati*	*Marathi*
Only Gujarati or Marathi	34.5	51.0	24.0	29.0	20.0	35.0
Only English	29.0	8.5	25.0	13.5	9.5	6.5
Varied Indian Languages*	1.5	4.0	2.5	4.5	5.5	6.0
English and Indian languages	32.5	28.0	13.5	10.5	6.0	7.5
Non-Readers	2.5	8.5	35.0	42.5	59.0	45.0

* Mainly Hindi, Gujarati, Marathi.

Table 2: Percentage Distribution of Language Preferences of Audio-Visual Media by Mother tongue

Language Preferences for Audio-Visual Media	*Films*		*Plays*		*Television*	
	Gujarati	*Marathi*	*Gujarati*	*Marathi*	*Gujarati*	*Marathi*
Only Gujarati or Marathi	0.5	7.5	41.0	53.0	2.0	20.5
Only Hindi	44.5	20.0	0.0	0.0	59.5	35.5
Only English	8.5	3.0	2.5	1.5	10.5	4.5
Varied Indian Languages	5.5	38.5	4.0	3.0	10.5	26.0
English and Indian languages	37.0	28.5	5.0	1.5	17.5	14.0
Non-viewers	4.0	2.5	47.5	41.0	0.0	0.0

Table 3: Percentage Distribution of English Language Media Usage

	News-papers	*Maga-zines*	*Books*	*Plays*	*Films*	*Tele-vision*	*Inter-net**
Non-user	5.5	38.8	52.0	44.3	3.3	0.0	66.8
Only Indian	45.5	30.0	33.3	50.5	58.3	77.0	—
English and Indian	30.3	12.0	6.8	3.3	32.8	15.5	—
Only English	18.8	19.3	8.0	2.0	5.8	7.5	33.3
	100.0	100.0	100.0	100.0	100.0	100.0	100.0

* it was assumed that Internet was used only in English.

Table 4: Percentage Distribution of Nature of Convergence Media Ownership and Usage by Mother tongue

Computer ownership	*Non-Users*	*Uses of Internet*			*Total*
		Mail/ chat	*Work/ research*	*Every-thing*	
No Computer	78.9	47.8	27.8	20.0	62.8
Only Computer	16.9	21.7	38.9	32.9	21.3
Computer with Internet Connection	4.1	30.4	33.3	47.1	16.0

Table 5: Percentage Distribution of Programme Preferences

Pro-gra-mme Pre-feren-ces	*News-cur-rent affairs*	*Ser-ials*	*Film/ film-based*	*Spo-rts*	*Music*	*Reli-gious*	*Car-toons*	*Horror shows*	*Game shows*	*Wo-men's Prog-ramme*	*Total N= 400*
Ist	26.8	52.3	4.5	8.8	4.8	0.8	1.0	0.3	1.0	0.0	100
IInd	22.3	20.4	27.7	10.5	11.3	4.0	1.1	0.3	1.9	0.5	93.0
IIIrd	17.5	8.8	30.5	17.9	17.2	7.5	0.3	0.0	0.3	0.0	77.0
IVth	19.7	6.9	12.4	26.6	21.6	10.1	1.4	0.0	0.9	0.5	54.5
Vth	11.7	8.1	6.3	13.5	42.3	15.3	1.8	0.0	0.0	0.9	27.8

Table 6: Percentage Distribution of Programme Preferences by Mother tongue

Preferred Programme Types	*First Preference*			*Second Preference*			*Third Preference*		
	Gujarati	*Marathi*	*Total*	*Gujarati*	*Marathi*	*Total*	*Gujarati*	*Marathi*	*Total*
News and Current Affairs	26.5	27.0	26.8	25.7	18.8	22.3	19.0	16.0	17.5
Serials	54.5	50.0	52.3	17.8	23.2	20.4	7.6	10.0	8.8
Film/Film-based	3.0	6.0	4.5	25.7	29.8	27.7	25.9	35.3	30.5
Sports	8.0	9.5	8.8	11.0	9.9	10.5	17.1	18.7	17.9
Music	4.5	5.0	4.8	11.5	11.0	11.3	17.1	17.3	17.2
Religious	1.0	0.5	0.8	5.2	2.8	4.0	12.0	2.7	7.5
Cartoons	1.5	0.5	1.0	1.6	0.6	1.1	0.6	0.0	0.3
Women's Programme	0.5	0.0	0.3	.5	0.6	0.5	0.0	0.0	0.0
Horror Shows	0.0	0.0	0.0	0.0	0.6	0.3	0.0	0.0	0.0
Game Shows	0.5	1.5	0.0	1.0	2.8	1.9	0.6	0.0	0.3
Total	100.0	100.0	100.0	100.0	100.0	100.0	100.0	100.0	100.0

Table 7: Percentage Distribution of Regularly Watched Programmes by Sex

Regularly watched programmes	*Sex*		*Total*
	Female	*Male*	
News and Current Affairs	5.5	16.5	11.0
Informative Programming*	5.0	7.5	6.3
Serials and Game Shows	54.0	36.0	45.0
Film/Music	5.0	6.5	5.8
No Specific Shows	12.0	22.5	17.3
English Serials/Films	8.0	5.5	6.8
Marathi Programmes	10.5	5.5	8.0
Total	100.0	100.0	100.0

* Women's Programmes, Cookery shows, National Geographic or Discovery channels, Quiz, Religious programmes, Yoga etc. and Sports.

Table 8: Percentage Distribution of Regularly Watched Programmes by Age

Regularly watched programmes	*Age*			*Total*
	20-30	30-40	40-50	
News and Current Affairs	8.4	17.2	10.2	11.0
Informative Programming	3.0	10.1	9.2	6.3
Serials and Game Shows	43.3	42.4	51.0	45.0
Film/Music	8.9	2.0	3.1	5.8
No Specific Shows	18.7	17.2	14.3	17.3
English Serials/Films	8.4	6.1	4.1	6.8
Marathi Programmes	9.4	5.1	8.2	8.0
Total	100.0	100.0	100.0	100.0

Table 9: Percentage Distribution of Language of TV Viewing by Mother tongue

Language of TV programmes	*Mother tongue*		*Total*
	Gujarati	*Marathi*	
Only Indian languages	72.0	82.0	77.0
Only English language	10.5	4.5	7.5
English and Indian languages	17.5	13.5	15.5
Total	100.0	100.0	100.0

Table 10: Percentage Distribution of Language Preferences for TV Programmes by Mother tongue

TV Language preferences	*Mother tongue*		*Total*
	Gujarati	*Marathi*	
Only Gujarati	1.5	0.0	0.8
Only Hindi	59.5	35.5	47.5
Only English	10.5	4.5	7.5
Only Marathi	0.5	20.5	10.5
Gujarati and Hindi	10.0	0.0	5.0
Hindi and English	15.5	2.5	9.0
Hindi and Marathi	0.5	26.0	13.3
English and Marathi	0.0	1.0	0.5
Gujarati, Hindi and English	0.5	0.0	0.3
Hindi, English and Marathi	0.5	9.5	5.0
All Four	1.0	0.5	0.8
Total	100.0	100.0	100.0

Bibliography

Abraham R. (1997) Cable guys take 'government' for a ride, *Sunday Observer*, 10-16 August 1997.

Acharya R.N. (1987) *Television in India:* Manas Publications, New Delhi.

Agrawal B.C. (1986) *Television Studies in India: The State of The Art:* Paper presented at International Television Studies Conference, UK, London, July 10-12, 1986.

———(1996) Transnational Television Expansion and Cultural Domination of Euro-American Life Style in Asia: Paper presented at PTC '96 at Honolulu, Hawaii, January 14-18, 1996.

———(1991) Passive Pleasures of Video: the technology intervention in the leisure of media rich: *ICCTR Journal*, III(1-2), 14-32.

Agrawal B.C. and Mira Aghi (1987) *Television and the Indian Child: A Handbook:* UNICEF, New Delhi.

Allen, Richard L. (1992) Communication Processes and Consumerism: The Case of Venezuela, *Howard Journal of Communications*, 4, 1-2, summer-fall, pp. 50-74, 1992.

Ang (1991) *Desperately seeking the audience*, Routledge, London.

Appadurai, A. (1996) *Modernity at Large: Cultural Dimensions of Globalisation*, University of Minnesota Press, Minneapolis.

Awasthy G.C. (1976) *Broadcasting in India:* Allied Publishers, Bombay.

Bajpai Shailaja (1997) *TV: Imbalanced Growth:* Vidura, Vol. 34, Issue-3, pp. 18-19, July-September, 1997.

Barber Benjamin (1996) *Jihad vs. Mc World*, Ballantine Books, New York.

Barker C. (2000) *Television, globalisation and cultural identities*, Open University Press, Buckingham.

Bearden William O., *et al.* (1993) *Handbook of Marketing Scales: Multi-Item Measures for Marketing and Consumer Behaviour Research*, Sage Publications, Newbury Park, CA.

Benerjee (2002) The local strike back? Media globalisation and localization in the new Asian Landscape, *Gazetteer*, Vol. 64(6), p. 518-535.

Bhat V.N. (1987) *Mind Managers and Defiled Doordarshan:* ABS Publications, Jalandhar.

Bhatt (1994) *Satellite Invasion of India:* Gyan Publishing House, New Delhi.

Bhattacharya V.R. (1976) *Communication in Global Village:* Chetana Publications, New Delhi.

Bock H.D. (1981) *Mass Communication Review Yearbook:* Vol. 2, Sage Publications, California.

Boyd-Barrett O. (1977) Media Imperialism: Towards an international framework for the analysis of media systems, p. 116-135 in Curran J. and Gurevitch M. (eds.) *Mass Communication and Society*, Edwards Arnold, London.

———(1998) Media Imperialism Reformulated, Ch. 9, p. 157-176 in Thussu D.K. (ed.) *Electronic Empires—Global Media and Local Resistance*, Arnold, London.

Briggs Susan (1998) Television in the Home and Family: In Smith and Peterson (ed.) *Television—An International History*: 1998, Ch. 8, pp. 109-121.

Brisil (1976) *Translation — Applications and Research*, Gardner Press, New York.

Britanica (1983) *The Encyclopaedia Britannica, Micropeadia, Ready Reference and Index*, Vol. V, Helen Hemingway Benton Publishers 1973-74, 1983—Fifteenth Edition.

Businessline (2002) Financial Daily from *The Hindu* group of publications, Tuesday, Feb 25, 2003, Businessline Internet edition retrieved from http://www.blonnet.com/2003/02/25/stories/2003022501790600.htm

Butcher, Melissa (2001) *The Dimensions of Cultural Change, Identity, Space and Transnational Television in India*: Doctorate Thesis, Centre for International Communication, Macquire University, Australia.

Butcher M. (2002), *When STAR Came to Town: Cultural Change and Transnational Television in India*, New Delhi: Sage.

Cable Waves (1998) *Directory of Indian Broadcasting Industry*, April 1998, Cable Waves Publishers Pvt. Ltd., New Delhi.

Cashmore-Ellis (1994) ... *and there was Television*, London and New York: Routledge.

Caughie John (1978) (ed.) *Television: Ideology and Exchange*: British Film Institute, Television Monograph 9, London.

Census (1981) Census of India 1981 Series 12 Maharashtra, Paper 1 of 1987: Household Population by Language mainly spoken in the Household, Table-HH16 part A (i).

Census (1991) Paper-2 of 1991, Series-1, Provisional Population Totals: Rural-Urban Districts and Table-29, Distribution of Population by Scheduled Language Groups: Page-121, Census of India, Maharashtra State District Profile, 1991.

———(2001) *Census of India 2001*, Provisional Population Totals, Paper-2 of 2001, Series-28, Maharashtra, pp. 60-66 .

Chadha and Kavoori (2000) Media Imperialism Revisited: Some findings from the Asian Case, pp. 415-433, *Media Culture and Society*, Vol. 22, No.4, July 2000.

Chakrabortty K. (2002) *Family in India*, Rawat Publishers, Jaipur.

Chandra V. (2000) *Television and Indian Culture*: Indian Publishers' & Distributors, Delhi

Chatterjee N. (1999) *AIDS-Related Information Exposure in the Mass Media and Discussion within Social Networks among Married Women in Bombay*, India, AIDS-Care 1999, Vol. 11, issue 4, August, p. 443-446. Cheung,-Chau-kiu; Chan,-Chi-fai (1996) Television Viewing and Mean World Value in Hong Kong's Adolescents, Social Behaviour and Personality; 1996, 24, 4, 351-364.

Chatterji P.C. (1987) *Broadcasting in India: Case Studies on Broadcasting Systems*: Sage in Association with International Institute of Communication, New Delhi.

Chaudhari K.K. (1986) *Maharashtra State Gazetteers Greater: Bombay District*, Volume 1, Gazetteers Department, Government of Maharashtra.

Chauhan K. (2001) *Television and Social Transformation: A Research in Mass Communication*, Sarup and Sons, New Delhi.

Cherry Colin (1971) *World Communication: Threat or Promise? A Socio-technical Approach*, Wiley Interscience, John Wiley, London.

Cheung ck. and Chang ch. fai (1996) Television Viewing and mean world value in Hong Kong's adolescents, *Social Behaviour and Personality*, 24, 4, 351-364.

Comstock G. (1980) Television and its Viewers what Eighties sees: In Wilhoit and Bock *Mass Communication Year Book*: Vol. 1, 1980, Ch. 29, pp. 491-508.

Comstock G. and Erica Scharret (1999) *Television—What's on, Who's watching and what it means:* Academic Press, New York.

Cunningham S., J. Sinclair and E. Jacka (1996) (eds.) New Patterns in Global Television-Peipheral Vision, OUP, Oxford.

Curran J. and M. Park (2000) *De-westernizing Media Studies*, Routledge, London.

Datta V. and Alwe M. (1999) Impact of Satellite television on family interaction patterns, *Contemporary Social Work*, Vol. XVI, October 1999, p. 87-91.

Dayan D. (2001) The peculiar public of television, *Media Culture and Society*, Vol. 23, p. 743-765, Sage, New Delhi.

DECU (1997) *Pursuit—Jhabua Development Communication Project*: Development and Educational Communication Unit, Indian Space Research Organization, Ahmedabad.

———(1998) *Interact—Training and Development Communication Channel*: Development and Educational Communication Unit, Indian Space Research Organization, Ahmedabad.

———(2000) *The Journey—25 years of satellite broadcasting in India*: Development and Educational Communication Unit, Indian Space Research Organization, Ahmedabad.

Deodhar P.S. (1991) *The Third Parent—Growth and Development of Indian Electronic Media*, Vikas Publishing House, New Delhi

Desai M. (2001) *Television Studies in India: Perspectives, Practices and Problems,* Ch 8, p. 106-130. In Chitty N. (2003) (ed.), Faces of Globalisation, Ganga Kaveri Publishing House, Varanasi.

Dhruva V. (2004) *Influence of television serials on women in urban slums of Mumbai,* Unpublished Masters Dissertation, Department of Extension Education, S.N.D.T. Women's University, Mumbai.

Dixit Kunda (1998) Broadcasting Developments—Impact on national and traditional culture, Media Asia, Vol. 25, No. 2, pp. 95, *Viewpoint.*

Dong Qingwen, Tan Alexis and Cao Xiaobing (1998) Socialization Effects of American Television and Movies in China, *Critical Studies;* 1998, 12, 311-327.

Doordarshan (1994) *Doordarshan 1994:* Audience Research Unit, Directorate General Doordarshan, New Delhi

———(1997) *Doordarshan 1997:*Audience Research Unit, Directorate General Doordarshan, New Delhi

———(1999) *Doordarshan 1999,* Audience Research Unit, Directorate General Doordarshan, New Delhi

———(2000) *Doordarshan 2000:* Audience Research Unit, Directorate General Doordarshan, New Delhi

———(2002) *Doordarshan India 2002:* Audience Research Unit, Prasar Bharati Broadcasting of India, Directorate General, Doordarshan, New Delhi

Doyle (1992) *The Future of Television—A Global Programming, Advertising, Technology and Growth,* NTC Publishing Group, Illinois

Dube C. (1990) *Tradition and Development:* Vikas Publishing House Pvt. Ltd., New Delhi.

Dunn-Robert-G.(1986)Television, Consumption and the Commodity Form, *Theory Culture and Society;* 1986, 3, 1, 49-64.

Eckersley Richard (1999) *Measuring Well-being: Material Progress and Quality of Life,* Keynote address at ACOSS Conference Nov 99 at the Australian National University, http://www.communitybuilders.nsw.gov.au/

Eko (2001) Jerry Springster and the Marlbro Man in Africa: Globalisation and Cultural Edecticism, *Ecquid Novi,* 2001, 22(1), p. 25-40.

Encyclopaedia (1959) *The Encyclopaedia Americana,* The International Reference Book, American Corporation, New York, 1959 edition, Vol. 15.

Encyclopaedia (1992) *Encyclopaedia of Americana*—International Edition, Groller Incorporated, Danbury, Vol. 15.

Fejes F. (1982) Media Imperialism: An assessment in Whitney, Wartella and Windahl (ed.) *Mass Communication Review Yearbook,* Vol. 3, Ch. 18, p. 345-353, Sage, Beverly Hills.

Fernandes L. (2000) Nationalizing 'the global': media images, cultural politics and the middle class in India. *Media, Culture and Society,* Vol. 22, issue 5, pp. 611-628, Sage.

Financial Express (2002) *DD Reach Tops Newspapers,* C&S Channels: IRS, Front Page, September 26, 2002.

Firoz M.D. (1998) Indian Cultural tradition under Western Eyes: STAR TV and otherness: *Social Action,* Vol. 48, No. 3, pp. 290-301.

Fiske J. (1987) *Television Culture,* Routledge, London.

Fournier Susan; Richins, Marsha-L. (1991) Some Theoretical and Popular Notions concerning Materialism, *Journal of Social Behaviour and Personality;* 1991, 6, 6 (special issue), pp. 403-414, 1991.

French D. and R. Michael (2000) (ed.) *Television in Contemporary Asia,* Sage, London.

French D. and R. Michael (1996) (ed.) *Contemporary Television—Eastern Perspectives:* Thousand Oaks, Sage, London.

Ghose, S. and A. Jahangirdar (1998) Trendy Conservatives: Cool in Kaliyug, *Outlook,* 1-14 December, pp. 61-70.

Golding and Harris (ed.) (1997) *Beyond Cultural Imperialism: Globalisation, Communication and New International Order,* Sage, London.

Golding P. (1998) Global Village or Cultural Pillage? The Unequal Inheritance of the Communication Revolution in R.W. McChesney, E.M. Wood, and J.B. Foster, (eds.) *Capitalism and the Information Age: The Political Economy of the Global Communication Revolution*, pp. 69-86, Monthly Review Press, New York.

Goldman R. (1992) *Reading Ads Socially*, Routledge, London.

Goonasekera (1987) The influence of television on cultural values with special reference to third world countries, *Media Asia*, Vol. 14, No.1, pp. 7-12.

Goonasekera A. and Paul S. (1998) *TV Without Borders—Asia Speaks Out* Singapore, Asian Media Information and Communication Centre.

Goregenli-Melek (1997) Individualist-Collectivist Tendencies in a Turkish Sample, *Journal of Cross Cultural Psychology;* 1997, 28, 6, Nov. 787-794.

Government of India (1963) *India 1963:* Publications Division, Ministry of Information and Broadcasting, New Delhi

———(1995) *Mass Media in India:* Publications Division, Ministry of Information and Broadcasting, New Delhi.

———Government of India (1998) *India 1998:* A reference manual, Ministry of Information and Broadcasting, February 1998.

———Government of India (1999) *Bilingualism and Trilingualism*, Census of India, Publications Division, New Delhi, 1991.

———Government of India (2001) *Mass Media in India 2001*, Ministry of Information and Broadcasting, Research, Reference and Training Division, New Delhi.

Greeson Larry E. and Williams, Rose Ann (1986) Social Implications of Music Videos for Youth: An Analysis of the Content and Effects of MTV, *Youth and Society;* 1986, 18, 2, Dec., pp. 177-189.

Grewal Inderpal (1999) Travelling Barbie: Indian Transnationality and New Consumer Subjects, *Positions*, Vol. 7, Issue 3, Winter, pp. 799-826.

Griffin Michael (2002) From Cultural Imperialism to Transnational Commercialisation: Shifting Paradigms in International Media Studies, *Global Media Journal*, Fall 2002, Vol.1, issue 1, http://www.lass.calumet.purdue.edu/From Cultural Imperialism.htm

Grossberge I. (1981) Interpreting the "crisis" of culture in Communication Theory, Ch. 5, pp. 147-159 In *Bock Mass Communication Review Yearbook:* vol. 2, Sage Publications, California.

Guback Thomas and Tapio Varis (1982) (ed.) *Transnational Communication and Cultural Industries,* Reports and Papers on Mass Communication No. 92, UNESCO, Paris.

Gupta N. (1998) *Switching Channels: Ideologies of Television in India,* Oxford University Press, New Delhi.

Gupta N. (2000) Just switch off!! Television: creating modern women, *Social Scientist,* Vol. 28, March/April 2000, pp. 322-323.

Gupta Varsha (1999) Show me the money: *The Economic Times,* July 14, 1999.

Hagen I. and J. Wasko (2000) (ed.) *Consuming Viewers? Production and Reception in Media Research,* Hampton, New Jersey.

Hali S.M. (2000) Media Imperialism, *Defence Journal of Armed Forces of Pakistan,* Opinion, June 2000. http://www.defencejournal.com/2000/june/imperialism.htm

Hamelink C.J. (1994) *Trends in World Communication—on Disempowerment and Self-empowerment,* South Bound-Third World Network, Penang, p. 69

Hargreaves-Alec-G. and Mahdjoub-Dalila (1997) Satellite Television Viewing among Ethnic Minorities in France, *European Journal of Communication;* 1997, 12, 4, Dec., pp. 459-477.

Hearlod S. (1986) A Sythesis of 1043 effects of television on social behaviour, pp. 65-133, Vol. 1, In Comstock (ed) *Public Communication and Behaviour,* Academic Press, New York.

Herman E. and Robert McChesney (1997) *The Global Media—The new missionaries of corporate capitalism:* Madhyam Books, New Delhi.

Hofstede G. (1984) *Culture's Consequences: International difference in work-related values,* Sage, Beverly Hills.

Hagiwara *et al.* (1999) The reception of global TV in Asia: An assessment of its impact on local culture, *Asian Journal of Communication,* Vol. 9, No. , p. 1-19.

Hoijer-B. (1992) Social Psychological perspectives in reception analysis, Ch. 14, p. 166-183 in Skovmand M. and Schroder K.C. (ed.) *Media Cultures: Reappraising transnational media,* Routledge, London.

Hui H. and Triandis H. (1986) Individualism-collectivism—a research of cross-cultural researchers: *Journal of Cross-Cultural Psychology,* 17, 225-228.

Hussain S. (1998) Fresh Paints, Old Tints, Brand Equity, *The Economic Times,* p. 1, 9-15 September 1998.

Hybel S. and Dana Vlloth (1978) *Broadcasting: An introduction to radio and television,* Dvan Nostrand Company, New York.

Indian Express (1992) STAR TV poses no threat to Doordarshan, September 26, 1992.

James-Beverly (1995) Learning to Consume: An Ethnographic Research of Cultural Change in Hungary, *Critical Studies in Mass Communication,* 1995, 12, 3, Sept., pp. 287-305.

Jehoram H.C. (1983) (ed.) *Cable Television media and copyright law aspects,* Reports to an International Literacy and Artistic Association and Dutch Copyright Association Symposium at Amsterdam, May 16-20 1982, Kluwer Law and Taxation Publishers, Netherlands.

Jha R. (1992) *Women and the Indian print media—portrayal and performance*: Chankay Publications, New Delhi.

John Sinclair, Stuart Cunningham, and Elizabeth Jacka (ed.) (1996), *New Patterns in Global Television—Peripheral Vision,* Oxford, England: Oxford U. Press.

Johnson Kirk (2000) *Television and Social Change in Rural India: A Research of Two Mountain Villages in Western Maharashtra,* Sage, New Delhi.

Joseph A. and K. Sharma (1993) (ed.) *Whose News? The media and women's issues*: Sage Publications, New Delhi.

Joshi Ila (1991) *Women dimension to television-policy, programming and personnel*: Concept, New Delhi.

Joshi P.C. (1985) Communication and Nation Building II, *Main Stream* No. 30.

———(1989) *Culture, Communication and Social Change*: New Delhi, Vikas Publishing House Pvt. Ltd.

Joshi S.R. (1998) Transborder Television in India In Goonasekera *et al.* 1998, pp. 3-37, *ibid.*

Juluri Vamsee (2002) Music television and the invention of youth culture in India, *Television and new media*, Vol. 3, No. 4, November 2002, p. 367-386.

Kang J.G. and Michael Morgan (1990) Culture Clash: Impact of U.S. Television in Korea in Martin and Hebert (ed.) *Current issues in International Communication*, pp. 293-301, Longman, New York.

Kang Jong G., Kapoor Suraj and Wolfe Arnold S. (1995) The Impact of Television Viewing on the Values Orientations of Indian Students: An Individualist-Collectivist Approach, *Howard Journal of Communications*, Vol. 6, Issue 3, November, pp. 188-205.

Kang Jong G.; Perry Stephen D. and Kang Seok(1999) The Relation between Television Viewing and the Values Orientations of Japanese Students, *Mass Communication and Society*; 1999, 2, 3-4, summer-fall, pp. 147-161.

Katz E. and Liebes T. (1994) *The export of Meaning: Cross-cultural readings of* "Dallas", Oxford University Press, Oxford.

Katz E. and George Wedell (1977) *Broadcasting in the Third World: Promise and Performance*: Macmillan Press, London.

Kazdin Alan E. (2000) *Encyclopaedia of Psychology* (ed.) American Psychological Association, Oxford University Press, Washington.

Keller D. (1981) *Network Television and American Society: Introduction to critical theory of television* In Whitney *et al.* 1982, Ch-23, pp. 411-442.

———(1987) TV, ideology and emancipatory popular culture In Newcomb (ed.) *Television: The Critical View*, Oxford, New York, pp. 471-503.

Keniston Kenneth (2002) Can the Cultures of India Survive the Information Age?, http://web.mit.edu/~kken/Public/papers1/Culture%20Survive.htm

Kishore K. (1994) The advent of STAR TV in India: Emerging Policy issues, *Media Asia*, 21(2), 1994.

Kitayam S. *et al.* (1993) The collective construction of self-esteem: Implications for culture, self and emotion, cited in

Cooper C.R., Theories Linking Culture and Psychology: Universal and Community specific processes, Annual Review of Psychology, Annual 1998, accessed at www.fundsarticles.com. on March 21, 2001.

Kivikuru U. and Tapio Varis (ed.) *Approaches to international communication*, textbook for journalism education, Finnish National Commission for UNESCO No.35, Helsinki.

Kohn-Paul-M. and Smart,-Reginald-G. (1984)The Impact of Television Advertising on Alcohol Consumption: An Experiment, *Journal of Studies on Alcohol;* 1984, 45, 4, July, 295-301.

Kosambi M. (1995) British Bombay and Marathi Mumbai: Some nineteenth century perceptions, ch-1, p. 3-24 in Patel S. and Thorner A. (1995) (Ed.) *Bombay—Mosaic of modern culture,* Oxford University Press, Delhi, 1995.

Kottak C.P. (1990) *Prime time Society: An Anthropological Analysis of television and culture,* Belize, Wadsworth.

Krishanan P. and Dighe A. (1990) *Affirmation and Denial—Construction of Feminity on Indian Television*: New Delhi, Sage Publication.

Kumar K. (1998) *History of Television in India: Political Economy Perspective,* Ch-2, p. 19-46, in Melkote *et al.* (1998.)

———(1999) Mass Communication in India, 11th Revised edition, Jaico Publishing House, Mumbai.

Lee P.S. (1997) Towards the Theory of transborder television in Goonasekera A. and P.S. Lee (ed.) *TV without Borders—Asia Speaksout,* p. 274-286, AMIC, Singapore.

Lee C.C. (1980) *Media Imperialism Reconsidered: The Homogenizing of Television Culture,* Sage, London.

Lee P.S.N. (2000) Television and Global Culture: assessing role of Television in globalisation, Ch. 11, pp. 188-198 in Wang *et al.* 2000).

Lee Sang-Chul and Susan K. Joe (2000) Key issues in the Korean Television industry: Programmes and market structure, Ch. 8, p. 131-149 in French and Richards (2000), *ibid.*

Lull J. (1988) *World Families Watch Television,* Newsbury Park, Sage.

Luthra H.R. (1986) *Indian Broadcasting*, Publications Division, Ministry of Information and Broadcasting, Government of India, New Delhi.

MacQuail D. (1997) *Audience Analysis*, Sage, Thousand Oaks.

Madden R. (2001) *Individualism, Interdependence, and Interpersonal Conflict, Spintech*, March 12, 2001, www.spintechmag.com/0103/rm0301.html, accessed on June 5, 2001.

Madhvani M. (1998) Double Trouble—Think Again column: Brand Equity, *The Economic Times*, 9-15 September, 1998, p. 1

Mahajan Kamlesh (1996) *Portrayal of Women on Television in India*, Comparative Social Research, 1996, 2 (supplement), pp. 53-67.

Mahanta A. (1987) *Research of Human Culture—A Theoretical Approach*: Omson Publications, Gauwhati

Malhotra S. and Rogers E. (2000) Satellite Television and the New Indian Woman, *Gazette*, SAGE Publications, 2000/10/01, Vol. 62, Issue 5, pp. 407-429

Mankekar D.R. (1979) *Media and Third World*, New Delhi, Indian Institute of Mass Communication.

Mankekar P. (1999) *Screening Culture, viewing politics—Television, womanhood and nation in modern India*: Oxford University Press, New Delhi

Masani Mehra (1976) *Broadcasting and People*, National Book Trust, New Delhi

Maslog C.C. (1984) Communication Research in Asia: Impression and Institutions: *Media Asia*, Vol. 11, No 2, pp. 67-75.

Mattelart A. (1979) *Multinational corporations and control of culture*, Harvester Press, Brighton.

Maureen Gibbins (1994) Crowded Skies—Issues in Asia-Pacific Satellite Development: *Media Asia*, Vol. 21, No. 2, Culture, Ethnicity, *Satellite Communication*, pp. 92-95, 1998.

McChesney R.W. (1998) The Political Economy of Global Communication. In R.W. McChesney, E.M. Wood, and J.B. Foster, (eds.) *Capitalism and the Information Age: The Political Economy of the Global Communication Revolution*, pp. 1-26, Monthly Review Press, New York.

McDaniel D. (2002) *Electronic Tigers of Southeast Asia, The Politics of Media,* Technology and National Development, Iowa University Press, Iowa.

McDowell S.D. (1997) Globalisation and Policy Choice: television and audiovisual services in India: *Media,* Culture and Society, Vol. 19, pp. 151-172, Sage London.

McMillin D. (2001) Localizing the global: Television and hybrid programming in India, *International Journal of Cultural Studies,* Vol. 4, Issue 1, pp. 45-68.

McQuail D. (1997) *Audience Analysis,* Sage, Thousand Oaks.

Melkote S.R. *et al.* (1998) *International Satellite Broadcasting in South-Asia—Political, Economic and Cultural Implications* Lanham, University Press of America.

Meyer W.H. (1988) *Transnational media and third world development: The structure and impact of imperialism,* Contributions to the research of mass media and communication, No. 11, Greenwood Press, New York.

Mishra R. (1996) Evaluation of educational Television Programme with special reference to UGC Countrywide Classroom, *University News,* January 1, 1996, pp. 10-14.

Mitra Ananda (1989) Trends in Television Research in India: Past, Present and Future, Banglore, *ICCTR Journal,* Vol II, 1989, No. 1-2, pp. 83-100.

———(1993) TV and the Nation: Doordarshan's India, *Media Asia,* Vol. 20 No. 1, pp. 39-44.

Modelski G. (1979) *Transnational Corporations and world order: Readings in International Political Economy,* WH Freeman and Co., C.A. in Wang J. (1997).

Mohanty Jagannath (1998) (ed.) *Studies in Educational Broadcasting: Television and Radio,* Vol. 1. and 2 Deep and Deep Publications, New Delhi.

Monterio and Jayasankar (2000) Between Normal and the imaginary: the Spectator—Self, the Other and Satellite Television in India, Ch. 15, pp. 301-321 in Hagen and Wasko (ed.) *ibid.*

Moores-Shaun (1993) Satellite TV as Cultural Sign: Consumption, Embedding an Articulation, *Media Culture and Society;* 1993, 15, 4, Oct., 621-639.

Morgan Michael; Leggett Susan; Shanahan James (1999) Television and Family Values: Was Dan Quayle Right?, *Mass Communication and Society;* 1999, 2, 1-2, winter-spring, 47-63.

Morley D. (1986) *Family Television: Cultural Power and Domestic Leisure,* Comedia, London.

Mullick K.S. (1974) *Tangled Tapes: The Inside Story of Indian Broadcasting,* Sterling Publishers, Jullundhar.

Muppiddi S.R. (1999) *The Uses and Gratifications of DoorDarshan and Eenadu TV—A research of a regional Indian television audience,* Ph.D. Dissertation, School Bowling State University, Bowling Green.

Murray J.P. and Kippax Susan (1980) *Television's Impact on Children and Adults: International Perspectives on Theory and Research* In Wilhoit and Bock,1980, Ch. 32, pp. 582-638.

Narasimhan S. (1998) Some observations about consumerism, *Madhyam,* Vol. XI, No. 2, Issues in Communication, Culture and Development, pp. 34-35, Bangalore.

Narayanan A (1987) *The Impact of Television on Viewers (A Case Research of Bombay TV Viewers),* Bombay, Somaiya Publications Pvt. Ltd.

Ninan Shevanti (1998) History and Indian Broadcasting Reforms. In Price M.E. and Stefaan G. (ed.) *Broadcasting Reforms in India: Media Law from a Global Perspective,* Chapter 1, pp. 1-21, Oxford University Press, Calcutta.

Nord D.P. (1983) An Economic Perspective on formula in Popular Culture in Wartella E., D. Charles Whitney and Sven Windanl (ed.) *Mass Communication Review Year Book,* Vol. 4, Ch. 18, pp. 287-301, Sage, London.

Nordenstreng K. and Tapio Varis (1974) (ed.) *Television traffic—A one-way street? A survey and analysis of the international flow of television programme material,* Reports and Papers on Mass Communication No. 70, UNESCO, Paris.

Oliveira O.S. (1986) Satellite TV and Dependency: An Empirical Approach, *Gazette,* Vol. 38, pp. 127-145.

Oliveria (1993) Brazilian Soaps Outshine Hollywood: Is Cultural Imperialism Fading out? In Schiller and Nordenstreng (ed.), *Beyond National Sovereignty: International Communication in the 1990s,* Norwood, New Jersey.

Page D. and William Crawley (2001) *Satellite over South Asia—Broadcasting Culture and Public Interest:* Sage Publications, New Delhi.

Paik H. and Comstock G. (1994) The Effects of Television Violence on Anti-social behaviour—A Meta analysis: *Communication Research*, 21 (4), pp. 516-546.

Patel Ila (1984) *Representation of women in mass media:* Workers papers-95: Institute of Rural Management. 'In Addendum' of seminar on 'women in media' December 8-9-1997 orgnised by Maxmuller Bhavan and Sophia Polytechnic, Bombay.

Patel Ila (2000) Television in rural context: A Case research of Nariyar, *Journal of Communication Studies*, Vol. XVIII, No. 1-2, pp. 6 to 19.

Patel S. and Thorner A. (1995) (ed.) *Bombay—Mosaic of Modern Culture*, Oxford University Press, Delhi, 1995.

Pathania G. (1998) Responses to transnational television in a STAR struck land: Doordarshan and STAR TV in India, Ch. 4, pp. 61-79 in Melkote *et al.* (1998).

Philo G. (1999) MTV's global footprint in Browne R. and M. Fishwick (eds.) *Global Village: Dead or Alive?*: Bowling Green State University, Bowling Green, OH.

Postman Neil (1987) *Amusing Ourselves to Death—Public Discourse in the Age of Show Business*, New Society, Methuen.

Price M.E. and Stefaan Gverhulst (1998) *(eds.) Broadcasting Reforms in India—Media Law from a global perspective:* Oxford University Press, Calcutta.

Price Monroe E. (1996) *Television, the Public Sphere and National Identity*, Oxford, England: Clarendon.

Rajagopal, Arvind (1996)Mediating Modernity: Theorizing Reception in a Non-Western Society, Communication Review; 1996, 1, 4, pp. 441-469.

Rao and Raghavan (1996) *Social effects of Mass Media in India*, Gyan Publishers, New Delhi.

Rao S. and Melkote (1998) Viewing of Doordarshan by Cable Subscribers in Bangalore India: Is there a difference with non-subscribers?, Ch. 10, pp. 181-203 in Melkote *et al.* (1998).

Rastogi Satish (1998) (ed.) *Educational Technology for Distance Education:* Rawat Publications, Jaipur.

Reddi U. (1989) Media and Culture in Indian society: Conflict or Cooperation?, *Media, Culture and Society,* Vol. 11, No. 4, October 1989, pp. 395-413.

Reddi U.V. (1996) *Rip Van Winkle: A Story of Indian Television:* pp. 231-245, Chapter 12 In French and Michael *(ed.), Contemporary Television—Eastern Perspectives:* Thousand Oaks, Sage, London.

Reddi U.V. (2000) Transnational Broadcasting and Multiculturalism: pp. 69-88, Chapter-5 In French and Richards *(ed.) Television in Contemporary Asia,* Sage, London

Richards M. (2000) *Television in Philippines: Structure and Issues,* Ch. 15, pp. 385-393 in French and Richards (2000).

Roach C. (1997) Cultural imperialism and resistance in media theory and literary theory, *Media Culture and Society,* Vol. 19, p. 47-66.

Robertson R. (1994) Globalisation or Glocalisation, *Journal of International Communication,* Vol.1, Issue 1, June, pp 33-52.

Saksena (2000) Forty years of television in India, *Vidura,* Vol. 37, Issue-1, January-March, 2000.

Saksena Gopal (1996) *Television in India: Changes and Challenges:* Vikas Publishing House, New Delhi.

Saksena Gopal (2000) Forty Years of Television in India: *Vidura,* Vol. 37, Issue 1, Jan.-March 2000.

Sarkar Mita (1997) Families—A Global Scenario: *Journal of Indian Anthropological Society,* Vol. 32, No. 3, Nov. 1997, pp. 263-267.

Schiffman L.G. and Kanuk Leslie Lazar (1999) *Consumer Behaviour*: Sixth Edition, Prentice Hall of India, New Delhi, The influence of culture on consumer behaviour, Ch. 14, pp. 406-437

Schiller H.I. (1976) *Communication and Cultural Domination,* Armonk, International Arts and Science Press, New York.

Schiller H.I. (1993) Not yet the Post-Imperialist Era, p. 97-116 in Roach C. (ed.) *Communication and Culture in War and Peace,* Newbury Park, Sage.

Scrase Timothy J. (2002) Television, the Middle Classes and the Transformation of Cultural Identities in West Bengal, India, *Gazette*, Vol. 64, No. 4, pp. 323-342, Sage Publications.

Sengupta S. and K.T. Frith (1997) Multinational Corporation advertising and cultural imperialism: A Content Analysis of Indian Television Commercials, *Asian Journal of Communication*, Vol. 7, No. 1, p. 1-18.

Senstrup and Goonasekera (1994) The Television landscape of four Asian countries, *Media Asia*, Vol. 21, No. 4, pp. 224-227.

Sereberny Mohammadi A. (1997) *The Many Faces of Imperialism*, Ch. 3, p. 49-68 in Golding and Haris (ed.), *ibid.*

Sereberny Mohammadi A. (1991) The Global and the Local in International Communications, In Curran J. and M. Gurevitch (eds.), *Mass Media and Society*, Edward Arnold, London.

Servaes J. and Patchanee Malikhao (1998) *A critical examination of a UNESCO research on Television Flows in Europe and Asia*, pp. 273-293, Chapter 14 in Melkote *et al.*, 1998.

Servaes J. and Rico Lie (2001) Media versus globalisation and localization, *Media Development*, 3, 2001, pp. 19-24.

Shah A. (2002) Take TV more seriously, *The Indian Express*, 26th September, 2002.

Shah Amrita (1997) *Hype, Hypocrisy and Television in Urban India:* Vikas Publishing House, New Delhi.

Shah K. and Udeshi S. (2001) Out of the Box office, *The Economic Times*, 15-21 June 2001, p. 3.

Sharma (1999) Kitne Sach hai channelo ke sarvekshan ke dave, *Cable Quest*, Issue II, Vol. 1, 15th June to 14th July 1999, pp. 33-41.

Sharma S.K. and Jagdish Singh (1972) *Television in Agricultural Transformation:* Indian Council for Agricultural Research, New Delhi.

Shevanti N. (1998) History of Indian Broadcasting Reforms, Ch. 1, pp. 1-21 In Price M.E. and Stefaan Gverhulst (1998) *(eds.) Broadcasting Reforms in India—Media Law from a Global Perspective:* Oxford University Press, Calcutta

Shield P. and Sandeep M. (1996) Integration, the Indian state and STAR TV—Policy and Thoery issues, *Gazette* 58:1-24, 1996.

Shingi P. *et al.* (1982) *Television and Knowledge Gap Hypothesis,* Centre for Management in Agriculture, Indian Institute of Management, Ahmedabad.

Sinclair J. (1997) The Business of International Broadcasting: Cultural Bridges and Barriers, *Asian Journal of Communication,* Vol. 7, No. 1, pp. 137-155.

———(1998) Culture as a Market Force—Corporate, Strategies for Asian Skies: Chapter 11, pp. 207-225, In Melkote *et al., International Satellite Broadcasting in South-Asia—Political, Economic and Cultural Implications:* University Press of America, Lanham.

Singh K.S. (1998) *Anthropological Survey of India, India's Communities,* Vol. V, Oxford, New Delhi.

Singh R. (1992) Impact of Television on Indian Society: *Interaction* X(1), 1992, pp. 135-160.

Singhal A. Arvind and E. Rogers E.M. (2001) *India's Communication Revolution: from bullock carts to cyber marts,* Sage Publications, New Delhi.

———(1989) *India's Information Revolution:* Sage Publications, New Delhi.

Singhal *et al.* (1998) Cultural Transcendence as Alternative to Cultural Imperialism—Role of Pro-social Entertainment Television Prgrammes in Developing Countries. In Melkote *et al.*, 1998, pp. 249-265, Chapter 13.

Sinha Arbind (1996) Development Dilemmas for Indian Television: pp. 302-320, Chapter 16 in French and Michael (ed.) *Contemporary Television—Eastern Perspectives:* Thousand Oaks, Sage, London.

Sinha Tara (1992) *Television comes of Age:* Seminar, 390 (February), 27-28.

Sitaram K.S. (1991) *Culture and Communication—A World View:* Associated Printers, Mysore.

Sklair Leslie (1994) *Sociology of Global System,* Baltimore, MD: Johns Hopkins U. Press.

Smith and Peterson (ed.) (1998) *Television: An International History*: Second edition, Oxford Press, London.

Sonwalkar P. (2001) India: Making of little cultural/media imperialism?, *Gazette*, Vol. 63(6), pp. 505-519.

Spark C. (1998) Is there a global public sphere? In Thussu K. (ed.) *Electronic empires: Global media and local resistance*, Arnold, London, Part I, Ch. 6, p. 108-124.

Srinivas M.N. (1977) *Social Change in Modern India*: Orient Longman Limited, New Delhi

Srivastava J.S. (1986) Rural India is not in the Picture in Agrawal B.C. and Sinha A (ed.), *"SITE to INSAT"*, Concept Publishing Co., New Delhi.

Stilling Erik A. (1997) The Electronic Melting Pot Hypothesis: The Cultivation of Acculturation among Hispanics through Television Viewing, *Howard Journal of Communications*; 1997, 8, 1, Jan.-Mar., pp. 77-100.

Straubhaar Joseph (2000) Culture, Language and Social Class in Globalisation of Television, Ch. 12, pp. 199-224 in Wang *et al.*

Straubhaar Joseph, Consuelo Campbell and Kristina Cahoon (1995) *From National to Regional Cultures: The Five Cultures and Television Markets in NAFTA.*

Sui Nam Lee P. (1995) A case against the thesis of communication imperialism: The audience's response to foreign TV in Hong Kong, *Australian Journal of Communicatin*, 22, 63-81.

Tahir S.N. (2000) Globalisation and National Television Networks: An analysis of Pakistan Television, Ch. 7, p. 347-361 in French and Richards (2000).

The Economic Times (1998) *Fresh Paint, Old Tints*, p. 1, Brand Equity Feature, September 9-15, 1998.

The Times of India (1992) Cable TV Will not Hit DD Says Minister, October 19, 1992.

The Times of India (1999) *Dancing to Consumer's Tunes:* Entertainment and Leisure—A Times Special Report, *The Times of India*, September 11, 1999, p. 15.

Thomas A.O. (1999) Regulating access to transnational satellite television: Shifting policies in Northeast Asia, *Gazette*, Vol. 61 (3-4), pp. 243-254.

Thomas P.N. (1998) South Asia In Smith A. and Paterson (ed.) *Television: An International History*, Second Edition, London, Oxford University Press, Chapter 16, pp. 201-207.

Thussu D.K. (1997) Globalisation of the 'Infotainment' Industry, *Third World Quarterly*, Vol. 18, Issue 2, June, pp. 391-396.

———(1999) Privatizing the airwaves: the impact of globalization on broadcasting in India: *Media, Culture and Society*, Vol. 21, pp. 125-131.

———(2000) The Hinglish Hegemony: The impact of Western television on broadcasting in India, Ch. 14, p. 293-311, in French and Richards (2000).

———(2002) *Managing the Media in an Era of Round-the-Clock News: notes from India's first tele-war, Journalism Studies*, Routledge, Part of the Taylor and Francis Group, 2002/04/01,Vol. 3, issue-2, p. 203-212

Tomlinson (1991) *Cultural Imperialism: A Critical introduction*, Pinter Publisher, London.

Tomlinson (1997) Cultural Globalization and cultural imperialism in Mohammadi (ed.) *International Communication and Globalization*, Sage Publications, London, p. 170-190.

Trager G.L. (1972) *Language and Languages*, Chandler Publishing Company.

Tremblay G. (1992) Is Quebec Culture doomed to become American?, *Canadian Journal of Communication*, Vol.17, No. 2, 1992.

Triandis H.C. (1996) The psychological measurement of cultural syndromes, *American Psychologist*, 51, 407-415.

Trivedi H.R. (1991) *Mass Media and New Horizons—Impact of TV and Video on Urban Millieu*: New Delhi, Concept Publishing Company.

Tunstall J. (1977) *The Media are American*, Columbia University Press, New York.

Tylor E.B. (1857) Primitive Culture in Mahanta A. (1987) *A Research of Human Culture—A Theoretical Approach*: Omson Publications, Gauwhati.

UN (1986) *Population Policy Paper No.6*, United Nations cited in Census, 2001, page-60

UNESCO (1953) *Television: A World Survey*, Report on the Facilities of Mass Communication, UNESCO, Paris.

———(1964) *World communications: Press, Radio, Television and Film*, Fourth Edition, UNESCO, Netherlands.

———(1971) *Broadcasting from space*, Reports and Papers on Mass Communication No.60, UNESCO, Paris.

———(1976) *Cross-cultural Broadcasting*, Reports and Papers on Mass Communication, No. 77, UNESCO, Paris.

———(1980) *Many Voices, One World: Towards a new just and more efficient world information and communication order*, MacBridge Commission Report, UNESCO, Paris.

———(1982) Guback Thomas and Tapio Varis (ed.), *Transnational Communication and Cultural Industries*, Reports and Papers on mass communication, No.92, UNESCO, Paris.

———(1994) *TV Transnationalisation: Europe and Asia*, Papers on mass communication No.103, Edited and abridged from reports by P. Sepstrup and A. Goonasekara, UNESCO, Paris.

———(2000) *World Culture Report, Cultural Diversity, Conflict and Pluralism*, UNESCO, Paris.

Variety (1976) and Variety (1978) January 7 and January 4 cited in Katze and George Wedell (1977) Broadcasting in the Third World: Promise and Performance, MacMillan Press, London, p. 97.

Varghese N. (2003) CAS may not deliver goods, Says Research, Businessline, *Sunday*, May 10, 2003 retrieved from http://thehindubusinessline.com/2003/05/11/stories/2003051101770100.htm

Varis T. (1984) The international flow of television programmes, *Journal of Communication*, 34, 143-152.

———(1985) *International flow of television programmes*, Reports and Papers on mass communication No.100, UNESCO, Paris.

———(1986) Global Problems and Communication, p. 7-27 In UNESCO (1994) *TV Transnationalisation: Europe and Asia*, Papers on mass communication No.103, Edited and abridged from reports by P. Sepstrup and A. Goonasekara, UNESCO, Paris.

Varma A. (2000) Impact of Watching International Television programmes on Adolescents in India—A Research Note: *Journal of Comparative Family Studies*, 31(1), Winter, pp. 117-126.

Varma S.J. (2000) Cable Manners, The Special Report, *Sunday Mid Day*, September 3, 2000, pp. 14-15.

Vasudeva S. and P. Chakravarty (1989) The epistemology of Indian mass communication Research: *Media, Culture and Society*, Vol 11 (1989, p. 415-433

Venkatesh Alladi (1995) Ethnoconsumerism: A New Paradigm to Research Cultural and Cross-Cultural Consumer Behaviour, in Costa, Janeen Arnold, and Bamossy, Gary J. (Eds.) *Marketing in a Multicultural World: Ethnicity, Nationalism and Cultural Identity*, Thousand Oaks, CA: Sage Publications, Inc, pp. 26-67.

Verma Suman and Reed W. Larson (2002) Television in Indian Adolescents' Lives: A Member of the Family, *Journal of Youth and Adolescence*, Volume 31, Number 3 , June 2002, pages: 177-183, Springer Netherlands.

Vijayanunni M. (1999) The Bilingual Scenario in India, *The Hindu*, 16 July, 1999.

Vilanilam J.V. (1996) The Socio-Cultural Dynamics of Indian Television: From SITE to Insight to Privatisation Ch 4, pp. 61-90 In French D. and R. Michael *Contemporary Television: Eastern Perspective*, Thousand Oaks, Sage, London.

Vilanilam, J. (1989) Television Advertising and the Indian Poor, *Media, Culture and Society*, 11, p. 485-97.

Wang G., Jan Servaes and A. Goonasekera (2000) *The Communications Landscape, Demystifying Media Globalisation*, Routledge, London.

Want G. and J. Servaes (2000) Introduction in Wang *et al.* (ed.), *ibid.*

Wang J. (1997) Global Media and Cultural Change: *Media Asia*, Vol. 24, N-1, pp. 15-22.

Ware W. and Michel Dupagne (1994) Effects of U.S. Television Programmes on Foreign Viewers: A Meta Analysis, *Journalism Quarterly*, Vol. 71, No. 4, Winter 1994, pp. 947-959.

Wei Ran and Pan Zhongdang (1999) Mass Media and Consumerist Values in the People's Republic of China, *International Journal of Public Opinion Research;* 1999, 11, 1, Spring, 75-96.

Wells A.F. (1972) *Picture tube imperialism? The impact of US Television on Latin America,* Maryknoll, New York, p. 137-143.

White Livingston A. (2000) *Reconsidering cultural imperialism theory* http://www.tbsjournal.com/Archives/Spring01/white.html Peer Reviewed Paper Competition Winner: Global Fusion 2000, Issue No. 4, Spring 2000.

Whitney, Wartella and Windahl (1982) *Mass Communication Review Yearbook :* Vol. 3, Sage Publications, California

Wilhoit and Bock (1980) *Mass Communication Review Yearbook:* Vol. 1 Sage Publication, California

Willnat L.; He Z.; Takeshita T.; López-Escobar E. (2002) Perceptions of Foreign Media Influence in Asia and Europe: The Third-Person Effect and Media Imperialism, *International Journal of Public Opinion Research,* June 2002, Vol.14, No. 2, pp. 175-192(18).

Wober M. and Gunter B. (1988) *Television and Social Control,* Avenbury, Aldershot.

Yadawa (1992) *Open Skies for TV,* Survey conducted by Dr. J.S. Yadawa, Director IIMC, September-October 1992 quoted in Bhatta (1994).

Zhao-Bin and Murdock Graham (1996) Young Pioneers: Children and the Making of Chinese Consumerism, *Cultural Studies;* 1996, 10, 2, May, pp. 201-217.

Websites

http://www.indiantelevision.com
http://www.thchoot.org
http://www.trai.gov.in

Index